HOW TO RAISE
CHICKENS

EVERYTHING YOU NEED TO KNOW

BREED GUIDE & SELECTION
PROPER CARE & HEALTHY FEEDING
BUILDING FACILITIES AND FENCING
SHOWING ADVICE

Christine Heinrichs

Voyageur Press

First published in 2007 by Voyageur Press, an imprint of MBI Publishing Company, 400 First Avenue North, Suite 300, Minneapolis, MN 55401 USA

Voyageur Press titles are also available at discounts in bulk quantity for industrial or sales-promotional use. For details write to Special Sales Manager at MBI Publishing Company, 400 First Avenue North, Suite 300, Minneapolis, MN 55401 USA.

To find out more about our books, join us online at www.VoyageurPress.com.

Library of Congress Cataloging-in-Publication Data

Heinrichs, Christine.
  How to raise chickens : everything you need to know / by Christine Heinrichs.
    p. cm.
  Includes index.
  ISBN-13: 978-0-7603-2828-6
  1. Chickens. I. Title.
SF487.H45 2007
636.5—dc22
                                        2006029850

Edited by Danielle J. Ibister
Designed by Jennifer Bergstrom

Printed in China

On the cover:
Photographs © Lynn M. Stone

On the back cover:
Top: Photo © Corallina Breuer
Middle: Photo © My Pet Chicken LLC
Bottom: Photo © Corallina Breuer

CONTENTS

DEDICATION
To Gordon and Nicole

ACKNOWLEDGMENTS

. . . .

To Nicole, who introduced me to chickens and woke me up to my life, and to Gordon, who made it all possible.

Writing this book was a gift that blessed my life every day. The opportunity to spend my time writing about chickens was one I had long hoped for. Chicken people are the most interesting and helpful group of people I have encountered. A phone call from a chicken person always brightens my day. I am grateful to all who called me with comments, ideas, suggestions, and problems. Every one of them enriched me.

My daughter and I acquired our first chickens back in the 1980s. Somehow along the way I joined the Society for the Preservation of Poultry Antiquities (SPPA). I became fascinated with the history of chickens and the dire future many historic breeds faced. Pretty soon, I was doing my part as publicity director.

That role gave me the chance to write about chickens and rare breeds. I could follow my own interests in my free time, while I worked on other people's subjects for pay during my working hours. Writing about rare breeds and the issues surrounding them was the best possible education I could have had. Craig Russell, who has raised chickens all his life and been involved with SPPA for many years, has been a patient teacher. The hours we spent on the telephone were rich and rewarding. His role in writing this book cannot be overestimated. He was in every way my cowriter.

Other SPPA officers and members have befriended me over the years. Ed Hart, another writer, provided personal and professional encouragement. Glenn Drowns and his wife, Linda, longtime secretary treasurer, handled the paperwork and finances under demanding circumstances. Dr. Charles Everett, who succeeded Mr. Drowns, has offered warmth along with his solid knowledge

of chickens. A Baptist minister, he has often provided spiritual guidance as well.

Monte Bowen is a true gentleman of the prairie. Hardworking and matter-of-fact, his work as a breeder has inspired me. His sense of humor has released me from taking myself too seriously more than once.

Don Cable has encouraged and supported my efforts. Mary Ann Harley has contributed her reliable good nature. Jeremy Trost has offered support, kind words, and his creative touch to matters both poultry and literary. Barry Koffler's website, FeatherSite, has been a resource and an inspiration. His comments have often straightened out my confusion. Jennifer Floyd shared her antique poultry library with me, copying many items so that I could refer to them. Horst Schmudde generously shared photos and lore about the Oriental game fowl on which he is the authority.

I have met many people along the way. Jim McLaughlin and Judith Kleinberg helped me on issues of small-flock processing, a bottleneck issue between small producers and consumers. Todd Wickstrom of Heritage Foods USA took time from his busy day to help me understand the role his company plays as a food broker connecting small producers with their customers.

Ron Kean, University of Wisconsin extension poultry specialist, answered so many questions for me. His enthusiasm for teaching and for poultry is an asset to the university's poultry program. I'm grateful he invited me to speak to his Poultry Consortium Center of Excellence class and join them for lunch afterward. Cherrie Nolden helped me with library presentations about chickens. She shared her practical knowledge of poultry-raising along with the knowledge she was gaining as a graduate student at the University of Wisconsin. Rob Porter, avian pathologist for the state of Wisconsin, shared information on helping small-flock owners for the health chapter. Harvey Ussery shared his experience with deep-litter management and hands-on caponizing.

Mary Zanoni is the founder of Farm for Life, a non-profit organization dedicated to "Farming in harmony with human life, physical and spiritual, with the life of our Earth, with the life of our fellow creatures." Her warmth and good humor cheered me on. Through our discussions of farming and politics, we found each other kindred souls.

K. J. Theodore graciously shared her medical and scientific knowledge, beyond the extensive material on her Shagbark bantam website. Dave Lewis answered my questions on breeding and culling. Nathan Larson guided me around the Troy Community Gardens, now home to three hens.

Elaine Belanger, editor of Backyard Poultry, helped provide an outlet for my chicken writings. We share a sense of humor that often lightened my day.

Ted Feitshans of North Carolina State University's Department of Agriculture and Resource Economics enlightened me on voluntary agricultural districts and other legal strategies to keep agricultural land in use. He went further and read and improved the chapter on that subject.

Karen Dunn, information services librarian at the University of Wisconsin's Steenbock Memorial Library, went far beyond her job to seek out references and help me learn the skills I needed to take advantage of them. Her patience and good humor are exceptional. Maia McNamara, reference librarian at the Madison Public Library, extended my first invitation to give a public presentation on chickens. It was unique for the library to echo with a rooster's crow. Her talent for connecting people helped me more than once. Stacey McKim explained digital images in terms even I could understand.

Writers need each other. Mary Bowman and the Madison Area Writers cheered me on. The Pinney Branch Writers, going over my words and helping me write more clearly, read more about chickens than any of them ever wanted.

Friends are invaluable. Susan McElhinney, Jan Gibbons-Ohr, and the sisters in my ward Relief Society listened patiently. Kay and Cheryl Nelson read chapters and commented, always tactfully.

As this project was getting started, I found a mentor through the Society of Professional Journalists in Ruth Ann Harnisch. Although chickens were outside her experience, she championed my cause and traveled with me to the conclusion. Our conversations buoyed me from one week to the next. I can never thank her enough.

She supported this work with a $1,000 donation to underwrite the photographs the book needed. The money helped defray expenses for Corallina Breuer, a young friend who already knows more about chickens than I ever will. Her knowledge informed her photography, resulting in charming and beautiful pictures that are also technically perfect. Corallina is looking forward to college and a career in poultry. I feel honored that she has accompanied me on this project. Some day I will claim, "I knew her when."

Others helped by sharing their photographs and digital images. Andrew Zimmerman, of the University of Florida's Department of Geological Sciences, sent images of chicken tractors. Traci Torres of My Pet Chicken came into my life at the moment when I needed her help, arriving on the scene with terrific pictures on her website.

With all the help I have received, it's still possible that errors have crept into the text. They are mine and cannot be blamed on any of these people who have so unselfishly helped me.

WHY KEEP CHICKENS?

Corallina Breuer

The pecking order organizes this flock eagerly waiting at the gate. One of the roosters is close to the front. Another looks over the hens. The senior hens take their places, ready to be the first out or at the center of any food that is delivered.

If you don't already own a small flock of chickens, you picked up this book because you want to. Everyone who lives in the country has considered raising chickens, and chickens are even moving back into cities and suburbia. The information compiled here will guide you to success, whether you have any experience or not.

Our urbanized society includes many members who continue to pursue interests in farming and rural life. 4-H has 6.8 million active members. Over 50 million Americans are 4-H alumni. The National FFA Organization, previously Future Farmers of America, has more than 490,000 members. Only 27 percent live in rural farm areas. The rest live in rural nonfarm areas (39 percent) and in cities and suburbs (34 percent).

Chickens are remarkably adaptable. Nearly every human culture except Eskimos has raised some kind of domestic chicken. Our fascination with chickens runs deep and primal.

Domestication likely began when an early human decided it was easier to weave a cage and put some baby birds in it than try to catch them or search for their nests. India's red jungle fowl, from which modern chickens are descended, are quick on the ground, clever about hiding their nests, and light enough to fly. Knocking them out of the trees with a slingshot must have taken exacting skills and a sharp eye.

Keeping chickens provided these early enthusiasts the opportunity to watch them—TV's precursor. Watching a pen of chickens is as mesmerizing as gazing into a fire. The phrase "pecking order" comes from chickens' social organization, and watching their interchicken jockeying is an entertaining pastime.

Like many birds, chickens naturally form flocks with a strong social hierarchy. Hens develop alliances and loyalties, and social life is rarely without incident. An extra worm to a low-ranking hen can set off a squabble or improve her standing. Life is never static in the chicken yard.

In our own chicken yard, a particularly unassertive hen that was last at everything, one day, discovered a large, juicy Jerusalem cricket for herself. She snatched it up in an unaccustomed gesture of self-assertion and took off, half a dozen hens in pursuit. They were outraged that she would presume to usurp this tasty delight for herself. Unwilling to release her prize, she flapped and ran until she had sufficient leeway behind a bush to gobble her treat in hasty triumph. After seeing her outwit her colleagues, I felt that perhaps her place at the bottom of the pecking order was not without its comforts.

MODERN POULTY KEEPING

Small-flock owners have many reasons for raising chickens. Some keep poultry only for their own use and enjoyment. Others raise production flocks for the growing organic and specialty egg and meat markets. A generation influenced by

Rachel Carson's *Silent Spring* understands the risks and limitations of the pesticides, antibiotics, and hormones that bless twentieth-century agricultural industrialization.

The post–World War II poultry industry replaced the economics of agriculture with those of business and industry. The focus moved from the flock and its economic place in the market to the financial equation of profit and loss: raw materials in, product out. Mass production conferred incomparable advantages over the small, independent poultry operation and the integrated farm. The business focus of industrial agriculture on profit trumps the integrated farm's concern for soil conservation, water quality, sustainable strategies, and environmentally sound practices such as the use of livestock wastes in crop rotation in addition to the economics of making a living.

Industrial producers strove for a single, ideal poultry unit that gave the most favorable conversion of feed into meat and eggs, the salable end products. Variety is the enemy of assembly-line production.

Color and variety of breeds lost value in huge poultry buildings that housed tens of thousands of nearly identical birds. Behavioral characteristics that stood out as

These Leghorns are generating antibodies, which are then naturally present in their eggs. The yolks will be dried and fed to other animals, giving them immunity to disease without having to take antibiotics. Eggs in modern commercial laying facilities are never touched by human hands.

Corallina Breuer

Larry Rana, USDA

Commercial poultry is raised in huge houses like this one, which may contain upward of twenty thousand broilers. The genetic similarity and crowded conditions make these birds vulnerable to rapid spread of disease. Waste disposal poses problems of contamination for ground water.

advantages in the small flock, such as curiosity and courage, became irrelevant or even undesirable. Broodiness, which interrupted egg-laying, was bred out of many breeds.

Applying industrial logic to livestock production reduced conditions to the minimum under which chickens could survive. As the national romance with modern technology waned late in the twentieth century, consumers began looking more closely at how and where their food was raised. Increased information about the chemicals fed to and applied to food animals raised concerns about effects on human health. Environmental issues were raised in response to the ammonia from commercial chicken barns and the disposal of waste. Consumers became concerned about humane living conditions for food animals.

Avian influenza recently captured headlines when a highly pathogenic strain, H5N1, emerged from the crowded conditions of poultry kept in Southeast Asia and China. Both the three-tiered system of small producers—which houses chickens on top, pigs in the middle, and fish

underneath, with droppings showering down to recycle as pig and fish feed—and mass commercial operations were implicated in the evolution of this strain, which infected and killed some poultry workers. Fear that it would mutate into a form transmissible between humans brought attention to poultry operations. Governments responded with mass culling to eliminate any possible transmission among chickens, creating havoc in rural societies that depend on chickens for food.

Gasoline prices topped three dollars a gallon, raising distribution costs. Concern for food safety increased, following a mad cow disease scare in beef and the growing awareness of vulnerability to terrorist attacks. The federal government introduced the National Animal Identification System (NAIS) in an effort to track food animals from birth to table.

Niche markets for gourmet organic and specialty foods are broadening. Consumers are more concerned with contaminants in their food, stimulating demand for

Troy Gardens

Troy Gardens is a community garden in Madison, Wisconsin. Three chickens joined the gardeners in 2006.

The chickens live in a coop built by Eagle Scout Toby Harrison-Noonan. Featuring wheels on one end, the portable coop can be moved around the gardens.

The 26-acre site is managed by the nonprofit group Friends of Troy Gardens and divided into 20-foot-by-20-foot garden plots, assigned each year to residents who apply for them.

Gardening tools, water, and supplies such as leaf mulch, compost, and manure are free. Some supplies, such as straw for mulch and floating row cover, are available at a small charge. Regular educational workshops are offered, at low or no cost, to help people learn to raise plants.

Plots are separated into sections for organic gardening, no-till gardening, and conventional gardening, where chemicals can be used. The garden discourages chemical use and organizes garden-wide integrated pest control.

Raised beds allow people with disabilities to participate. Although that section of the gardens is not yet fully wheelchair accessible, the Friends of Troy Gardens has plans for a fully accessible section.

The gardens have attracted the city's Hmong population. Garden education director Nathan Larson has supported their efforts by arranging plots so that clans can garden together.

An adjacent five acres is being developed as affordable housing on the co-housing model. Part of the 26 acres is being restored as natural area.

The gardens provide a common meeting ground for residents who might not otherwise get to know each other. The produce gardeners raise supports family nutrition and saves money, helping stretch limited incomes.

organic and locally raised food. Increased production is bringing these niche market prices down while mass-market foods struggle with cost increases. Farmers' markets have increased to more than three thousand. More than one thousand community-supported agriculture groups have connected buyers directly with producers. These groups offer a new model of food production, sales, and distribution. Customers buy into the farmer's budget, taking on a share of the risk in return for high-quality foods they are assured are raised with thoughtful methods. Small farmers and gardeners have created many variations to meet the needs of producer and consumer. Some sell subscriptions that provide weekly deliveries of produce.

Consumers, all of whom buy and eat food, have more confidence that small-scale farms produce safe, nutritious food. A recent Roper survey found 85 percent of a national sample trusted small farms for the quality of food, 70 percent for the safety of food, and 69 percent trusted them to produce it with techniques that don't harm the environment.

With big producers elbowing into the organic market, consumers' best bet for safe, nutritious food is to buy from local producers if they can't grow their own. Food brokers like Heritage Foods USA have emerged to connect small producers with consumers.

The public is looking closer at poultry. As people are getting to know more about chickens, they are regaining the affection that characterized the relationship throughout recorded history and, no doubt, before.

FARM CHICKENS

You might remember visiting a grandparent who kept chickens in the yard. You may have been frightened by a protective hen or an aggressive rooster, or you may recall the surprise of finding eggs. Until the middle of the twentieth century, raising a few chickens was common. Since everyone kept them, everyone knew about them: what they ate, when to expect eggs, what made them sick. The chickens lived in the yard and provided egg money.

Lore developed, some of it accurate and some more hopeful than useful. If you didn't know the answer to a question about your chickens, you could ask your neighbor or your mother or someone at church. It was an integral part of the culture.

As people moved to cities and suburban developments in the 1950s, the general public lost those poultry-handling and poultry-management skills. Much of it has become arcane, although it is invaluable to those who wish to raise small flocks.

Raising birds for meat has a different emphasis than raising exhibition birds, but many of the management practices are the same.

Small-flock owners rarely see their birds as narrowly as simply food or only show animals. The utility values of egg-laying and meat are integral parts of all breeds. After all, that was why they were traditionally kept. Small-flock owners naturally appreciate their chickens' beauty. Many find their interest in specialty breeds piqued. Those keeping birds for the market develop a desire to show. Showing provides far more than an opportunity to be recognized as

the best. It is a chance to meet others of like interest and find new birds, new bloodlines, and share ideas.

Recent publicity about especially rare breeds has resulted in bursts of interest, rejuvenating breeds that were nearly gone. Public interest and excitement about poultry are growing.

Small farms are on the increase, according to the most recent Census of Agriculture by the United States Department of Agriculture (USDA), up to more than 560,000. Many of those small farmers are retirees or second-career farmers. Of those who have been farming four years or less, 27 percent are over fifty-five years old.

These advocates are living out lifelong desires to farm, grow their own food, be self-sufficient, contribute to the local food economy, and feed the spiritual needs that rural life can provide. Chickens are part of that life.

BACKYARD CHICKENS

Cities and smaller communities are increasingly making it legal to raise chickens. A certain prejudice against chickens—a whiff of class division—can trigger opposition, but most communities find some way to allow chickens in urban and suburban settings.

"In cities, the chic is back in chickens," said Traci Torres, a New Jersey marketing director who left her job to start an online business of chicken-related merchandise, My Pet Chicken, LLC.

Ken Hammond, USDA

Left to right, Eva Mae Turner, Windsor Ingram, Peggy Booth, Eunice Daniels, and Dorothy Stone pose at the community garden at the Northpoint Apartments in Mt. Olive, North Carolina.

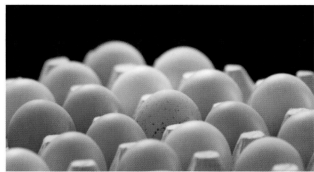

Courtesy of My Pet Chicken, LLC

Fresh eggs from local hens have a natural appeal that grocery-store eggs can't compete with. Studies indicate that eggs from hens that forage for part of their diet are more nutritious than commercial eggs.

Having truly fresh eggs from chickens of your personal acquaintance is one of the reasons people raise their own chickens today. Like so many other home-raised products, the flavor can't be compared to store-bought. The nutritional value reflects the hens' feeding, something you can influence if you raise your own.

Gathering eggs can feel like found money. Children delight in searching nest boxes and gathering eggs. One friend's daughter was going through a particularly picky-eater stage. Refusing to eat nearly everything, she discovered our eggs. That kid made those fresh eggs her mainstay for months, until she outgrew whatever it is that makes kids funny about food.

A dozen or so hens will provide any family and most of your neighbors with plenty of eggs. Sharing them with neighbors generates enormous goodwill. One Baltimore breeder maintains congenial relations with his neighbors, even though he keeps a rooster, through his generous gifts of eggs and his open-door policy. A sense of humor helps, too.

Small groups do better than one or two chickens. Chickens are highly social and need each other's companionship, although some singletons make do by befriending the cat. Remember that chickens are subject to predation and even the most careful caretaker loses some chickens, so keeping more than just a couple makes sense.

With a few more, you can have fresh meat as well. The accepted wisdom is that if you consider your birds meat,

Corallina Breuer

A good way to find fresh eggs is from a local poultry program. The University of Wisconsin sells eggs from its flocks on the honor system for 50 cents a dozen. Eating eggs are refrigerated at 40 degrees Fahrenheit; hatching eggs are kept separate at 60 degrees.

11

Christine Heinrichs

Community gardens like this one in Madison, Wisconsin, preserve green space in urban and suburban areas.

Christine Heinrichs

The community garden's chicken coop was built as an Eagle Scout project by Toby Harrison-Noonan. The two-by-four with a plastic bag hanging off it was used to carry the coop into its location in the garden. It can be used to move it to other locations.

Christine Heinrichs

Farmers' markets help small producers find their customers. The popularity of these markets is growing. Local food sources provide food system security because they do not rely on transporting food over long distances.

you will not want to name them. You will likely develop some favorites who are granted names and will be around for a long time anyway, perhaps even staying on as retirees after they aren't laying many eggs anymore. If you are serious about breeding, however, you will cull your flock and fill your freezer and pot with chicken tasty enough to spoil your palate for fast-food chicken.

Chickens make an accessible agricultural project for adults and youngsters alike. They are a lot easier to raise than cattle or pigs, and you get eggs every day. Parents without farm backgrounds are less intimidated by chickens. Several SPPA members began raising poultry because of some disability or physical limitation. Having your own flock of chickens gives you a strong foundation of self-reliance.

LIVESTOCK CONSERVATION

Genetic diversity is nature's insurance policy, as much in domestic populations as in wild ones. Historic livestock breeds are slipping into extinction the same way wild ones are.

Corallina Breuer

This pair of Old English Game bantams are living history. Old English Games, as their name suggests, are a traditional breed that came to America with the colonists. They remain popular among fanciers and are sturdy, productive birds.

Corallina Breuer

This Buff Catalana rooster struts across the yard with all the confidence of knowing the proud heritage of chickens. The Catalana is a Spanish breed gaining popularity again in the United States.

The diversity of chickens speaks to the diversity of people. They touch us in profound ways. Robert Frost captured that fascination and pride in his poem about his favorite chicken:

A Blue Ribbon at Amesbury

Such a fine pullet ought to go
All coiffured to a winter show,
And be exhibited, and win,
The answer is this one has been—

And come with all her honors home,
Her golden leg, her coral comb,
Her fluff of plumage, white as chalk,
Her style, were all the fancy's talk.

It seems as if you must have heard.
She scored an almost perfect bird.
In her we make ourselves acquainted
With one a Sewell might have painted.

Here common with the flock again,
At home in her abiding pen,
She lingers feeding at the trough,
The last to let night drive her off.

The one who gave her ankle-band,
Her keeper, empty pail in hand,
He lingers too, averse to slight
His chores for all the wintry night.

He leans against the dusty wall,
Immured almost beyond recall,
A depth past many swinging doors
And many litter-muffled floors.

He meditates the breeder's art.
He has a half a mind to start,
With her for Mother Eve, a race
That shall all living things displace.

'Tis ritual with her to lay
The full six days, then rest a day;
At which rate barring broodiness
She may well score an egg success.

The gatherer can always tell
Her well-turned egg's brown shapely shell,
As safe a vehicle of seed
As is vouchsafed to feathered breed.

No human specter at the feast
Can scant or hurry her the least.
She takes her time to take her fill.
She whets a sleepy sated bill.

She gropes across the pen alone
To peck herself a precious stone.
She waters at the patent fount
And so to roost, the last to mount.

The roost is her extent of flight,
Yet once she rises to the height,
She shoulders with a wing so strong
She makes the whole flock move along.

The night is setting in to blow.
It scours the windowpane with snow,
But barely gets from them or her
For comment a complacent chirr.

The lowly pen is yet a hold
Against the dark and wind and cold
To give a prospect to a plan
And warrant prudence in a man.

Current industrial agricultural practices have reduced genetic diversity in poultry populations to dangerous levels. Raising rare and historic breeds puts you on the front lines of practical conservation and living history.

The genetic uniformity of commercial poultry courts disaster with the birds' limited ability to respond to disease or environmental change. Traditional breeds perpetuate genetic resilience that may be needed in the future. As with wild species, we may not know what we need until it is gone.

Poultry was developed to meet the practical, religious, and aesthetic purposes of varied cultures. Feather color and pattern, comb size and shape, egg color and size, and unusual characteristics such as naked necks, long crows, and a fifth toe create endless variations. Learning the history of the breed that finds a special place in your life opens a door to understanding those who came before us.

QUALITY OF LIFE

People like chickens. When I first started keeping a few in my suburban backyard in San Jose, California, heartland of Silicon Valley, the first thing many visitors said was a wistful, "I always wanted to have chickens." When one chick unexpectedly grew up to crow, neighbors told me how charming it was to hear the sounds of the countryside.

Chicken owners are widely diverse. Some keep a few and make pets of them. They aren't cats or dogs but can be delightfully personable. Their individuality adds the same sparkle to our lives that other animal companions do.

Those with more businesslike reasons for raising chickens nevertheless enjoy their company and take pride in raising healthy, vigorous, beautiful chickens. They contribute to local economies and offer alternatives to industrialized agriculture.

Corallina Breuer

A rooster's crow brings back fond farm memories to some but irritates others beyond endurance. The frequency of crowing varies among roosters, but many crow a lot. This Ameraucana crossbred rooster is happy to tell the world his news.

Egg Color

Egg color is often correlated with earlobe color. The rule of thumb is that red earlobes equal brown eggs and white earlobes equal white eggs. But there many exceptions.

Among commercial egg-layers—dominated by Leghorns laying white eggs and Rhode Island Reds laying brown eggs—the correlation holds true. However, Penedescencas have white earlobes and lay especially dark brown eggs. Araucanas and Ameraucanas lay blue eggs but have red earlobes.

Because the traits are not linked genetically, they can and have been sorted separately in selective breeding through the years.

Corallina Breuer

This Black Cochin cock is reason enough to keep chickens. His glossy black feathers touch our aesthetic appreciation of beauty. His attention to detail makes him capable of foraging for part of his food, scratching up bugs and seeds.

Eggs

Hens lay eggs whether a rooster is around or not. Without a rooster, the eggs will not be fertile.

Egg-laying varies by breed, by season, and by individual. Laying is related to day length, extended in commercial operations where the chickens are inside all the time and exposed only to artificial light. Most breeds take time off in the winter, although some are particularly reliable about laying during the shorter days of cold months, such as the Canadian Chantecler, the Rhode Island Red, and the thickly feathered Wyandotte.

Hens lay fewer eggs as they age, with production declining gradually after three years of age.

Eggs vary from pure chalky white to dark chocolate brown, with Araucanas and their descendants laying blue and khaki green eggs.

Leghorns are the preferred commercial egg breed, but many traditional breeds are considered dual-purpose breeds: good egg layers that are also good table birds. Leghorns are considered too scrawny to be good eating.

A Black Australorp is reported to have laid a record 364 eggs in one year, but most good layers take a day or two off each week. Hens lay less during their annual molt, usually in late summer. Breeds that maintain the instinct to become broody will stop laying during that three-week period. Broodiness has been selectively bred out of many breeds and strains to avoid this eggless period.

Free-range hens that forage for part of their food, eating grass, seeds, and bugs, have been determined to lay more nutritious eggs than hens from commercial farms. *Mother Earth News* worked with Skaggs Nutrition Laboratory at Utah State University and Food Products Laboratory in Portland, Oregon, to test the eggs of free-range birds. The eggs contained up to twice as much vitamin E; up to six times as much beta carotene, a form of vitamin A; and four times as much omega-3 fatty acid. They had half the cholesterol of commercially raised eggs.

Hens are sociable and willing to work with others. Having different breeds poses no barrier to flocking your birds together, although sizes may make a difference. With enough space, chickens rarely become aggressive toward each other.

Courtesy of My Pet Chicken, LLC

Courtesy of the Garfield Farm Museum

Chickens are social, perching birds. They flock together naturally, as demonstrated by these Black and White Javas at Garfield Farm Museum in La Fox, Illinois. Several Narragansett turkeys join them without feeling out of place.

CHAPTER 2

• •

CHICKENS

AND

HISTORY

Birds are descended from dinosaurs. As with any developing scientific subject, some disputes remain as to exactly where and when things happened. But although 150 million years is a long time, the evidence is in: birds descended from dinosaurs.

When you observe your chickens, you see the obvious similarities: They lay eggs. They have scales on their legs, an S-shaped neck, and an erect stance with feet positioned directly below the body. Most have four-toed feet supported by three main toes.

Scientists also weigh more subtle and concealed evidence, like chickens' hollow bones, the number of vertebrae incorporated into the hip, and the way the clavicles fuse to form the furcula, or wishbone.

Feathers showed up on dinosaurs about 145 million years ago, on Archaeopteryx, a crow-sized bird fossil from Germany. Excavations in China are turning up fossils dating back 130 million years that show hairlike features that may be proto-feathers. As birds developed, they specialized to occupy environmental niches. Fowl, particularly ground-nesting birds, eventually led to the jungle fowl, *Gallus gallus*, of Southeast Asia, from which modern domestic chickens arose.

By the time humans got around to writing things down, chickens had been at their side for a long time.

DOMESTICATION

It seems intuitive that if early humans were throwing bones to friendly wolves and domesticating them as dogs ten thousand or twenty thousand years ago, they were likely picking up ground-nesting birds and bringing them back to the cave.

Some Upper Paleolithic sites in England and Greece suggest a species of *Gallus* may have lived through the Stone Age. The species could have become extinct later as a result of glaciation or hunting by early humans.

Domestication is a complex process, and not all animals are suited to it. Relatively few species have ever become domesticated. A species needs to be naturally inclined to respond to a dominance hierarchy, live in large gregarious social groups such as flocks, and readily adapt to humans.

The ability to reproduce in captivity is crucial to successful domestication. It allows humans to impose selective breeding and to shape the animal to greater human usefulness.

In the case of chickens, they became larger in domestication than their wild relatives. Small size and speed are advantages in wild ground-nesting bird populations. The small, wiry male is at a genetic competitive advantage. Natural variations in feather color occur. In the wild, colors would respond to evolutionary pressures, such as mating displays in the male and protective coloration in the female. In domestication, colors that would be maladaptive in the wild could be selectively bred. Instead of lunch, they become show birds.

Stephen Acosta was a longtime breeder who acquired jungle fowl from the wild in the 1920s. Although he bred them only to each other, by the 1970s they had doubled in size. Craig Russell, president of the Society for the Preservation of Poultry Antiquities (SPPA), visited Acosta and observed, "If you are not selecting for smaller size, they will get bigger fast."

BREED DEVELOPMENT

Chicken breeds developed distinct traits early on. At least four species of jungle fowl exist, the red jungle fowl being the main progenitor of today's domestic chicken. Land races, such as Games, Javas, and Dorkings, are the breeds that developed naturally. Other domesticated breeds were developed from them.

Sumatras are a land-race breed, which developed naturally in Southeast Asia. Domestic chickens developed from jungle fowl in India and Southeast Asia.

This jungle fowl/domestic chicken cross is typical of chickens that live wild in the South Sea Islands.

Composite breeds were developed by crossing and recrossing breeds over the years. Some of the composite breeds have a long history, but new composites are being developed all the time. Chicken breeds are not standing still.

Chickens were originally domesticated in Southeast Asia. Archaeological evidence shows that domesticated chickens were well established in China by 6000 BC. Chicken bones, along with dog and pig bones, were recovered from the Cishan site in Wu'an county of Hebei province. The number and size of the chicken bones suggest that they were from domesticated birds.

From China, they were introduced to Korea during the Yayoi Period (300 BC to AD 300), then to Japan. Japanese breeders, like all others, selectively bred chickens to develop their own specialties, including long-tailed breeds like the Onigadori.

Chickens were probably domesticated more than once. That is, the phenomenon of domestication wasn't the brainchild of a single individual or village that then diffused by trade or other cultural contact. The attraction of wild jungle fowl likely resulted in their captivity and domestication many times in many places.

Jungle fowl roosters are naturally territorial toward each other. Domestic birds may originally have been kept for entertainment and cockfighting. At times, in some places, chickens have been kept entirely for purposes other than food. Around 1000 BC in India, eating fowl was forbidden. White chickens in ancient Greece were sacred to Zeus, and Pythagoras forbid his followers from eating them.

By 2000 BC, chickens had been domesticated or acquired in India. At the Mohenjo-Daro archaeological site in the Indus Valley in present-day Pakistan, chicken bones were found in the upper levels of excavations. In the Harappan culture of Punjab, chickens were part of life from 2500 to 2100 BC.

JUNGLE FOWL TO CHICKENS

Jungle fowl share the characteristics of other animals that are capable of being domesticated: They offer some benefit to humans and they are able to live and reproduce in captivity.

The jungle fowl of today are classified into four species. Modern biochemical analysis indicates that these four made varying contributions to modern domestic chickens. *Gallus gallus*, the red jungle fowl, is considered the primary ancestor of modern domestic chickens, but Ceylon jungle fowl (*G. lafayettei*), gray jungle fowl (*G. sonnerati*), and green jungle fowl (*G. varius*) also show significant influence. Whether a fifth species, *G. giganteus*, also existed but is now extinct remains speculative but provocative.

All jungle fowl are social and live in flocks but with variations in social structure. Some birds form mated pairs for life, while some males maintain harems. The four extant species can be interbred but with varying reports as to fertility, reproduction being the crucial element. Some report the second generation does not survive or has sharply reduced fertility. This characteristic suggests the breeds are fairly well-separated genetically and over time. The geographic barriers of the Indian subcontinent and the Himalayas to the north establish natural limits to separate these species.

Red jungle fowl, also known as *G. bankiva* and *G. ferrugineus*, range today from Kashmir across Burma to Tonkin. The cocks are spectacularly colored. The single upright serrated red comb and single pair of wattles are distinguishing characteristics. Hens are drab colored, in keeping with most ground-nesting female birds. Both sexes display the black-breasted red plumage modern poultry breeders define on Black Breasted Red Games, Red Dorkings, and Light Brown Leghorns. They have dark, usually slate, legs.

Four kinds of jungle fowl still exist. The red jungle fowl contributed the most to domesticated chickens. This map shows the distribution of the four kinds of jungle fowl today.

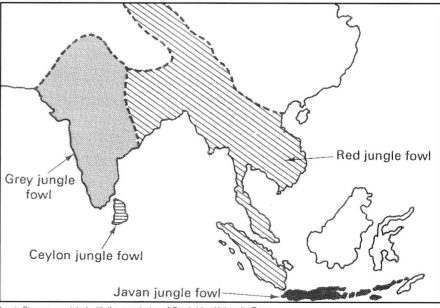

Lewis Stevens, reprinted with the permission of Cambridge University Press

Some consider red jungle fowl to have developed five subspecies that differ in ear-lobe color from white to red; shape and length of neck hackle feathers in roosters; shade of rooster plumage, which can range from golden yellow to mahogany; shade and sheen of plumage; comb or lack of comb in females; and degree of eclipse molt in males. In an eclipse molt, the male molts to the drab appearance of the hen for a while, then completes the molt and acquires his rooster's plumage again. These variations form distinct populations that may have resulted from natural breeding pressures or crossing with domestic birds.

Ceylon jungle fowl, isolated on the island now known as Sri Lanka, are usually given the scientific name G. *lafayetti* but are also known as G. *stanleyi* and have been called Ceylonese, Sinhalese, or red and yellow jungle fowl.

The roosters have developed different plumage. Their feathers are pointed and fringed, rather than rounded, and red where red jungle fowl are black. Some have a bluish-purple patch on the upper breast. Male Ceylons have a single serrated comb, bright yellow edged in brilliant crimson. Males are brighter and more orange than red jungle fowl. Females are lighter than reds. Legs are yellow. Craig Russell, of the SPPA, has crossed reds with Ceylons and found them not only fertile but exhibiting hybrid vigor. Other breeders have reported Ceylon-like individuals appearing spontaneously from red matings. Whether this reflects the origin of Ceylons within the red genome or previous red crossings is unknown.

Gray jungle fowl, classified as G. *sonnerati*, are native to the hill forests of western and southern India. They have "sealing wax" spangles on some feathers in the neck, hackle, saddle, and wing coverts. These spangles are actually flattened areas of the feather shaft, or rachis. Along the shaft, the spots are white; at the end, they are yellow and

Gallus giganteus

Some historians hypothesize that an additional fowl species, *Gallus giganteus* or *G. giganleus*, is the missing ancestor of domestic chickens that would have contributed traits more adaptive to a grassland environment than bush or jungle. A provocative bit of anatomy is one of the pieces of evidence suggesting this link.

The opening between the brain and spinal cord is horizontal in jungle fowl. In Oriental and Asiatic chickens, this opening is perpendicular. While it could be a random mutation, such a concealed trait would not be the result of selective breeding.

Internal ear structures apparently linked to cushion and pea combs also suggest a separate genetic history, without clear ancestry for the combs.

The distinctive tall, long-legged, and upright stance of Malayan fowl, such as Malay, Aseel, Shamo, Madagascar Game, and Saipan suggest that other influences may exist in the mist of the past. These breeds also have proportionately smaller wings. Their proportionately longer and more muscular legs are an adaptation that suggests running rather than flying, a behavior that would be associated with evolutionary success in a grassland environment.

Others put *G. giganteus* in the same category as Bigfoot. The differences in chicken breeds, while extreme, are no greater than the differences among dog breeds that have evolved since the domestication of wolves. Future research will surely add to our knowledge and help resolve this question.

shredded. The spangles feel smooth and waxy. Gray jungle fowl also have a single red serrated comb and a pair of red wattles. Their body feathers are black with white shafts and gray borders, and they have black wing and tail feathers. Hens have white breast feathers with broad black or brown penciling. They have red legs.

Gray jungle fowl are often monogamous and the cocks participate in raising the chicks. Wild-caught birds remain skittish, but chicks fostered by domestic birds become tame.

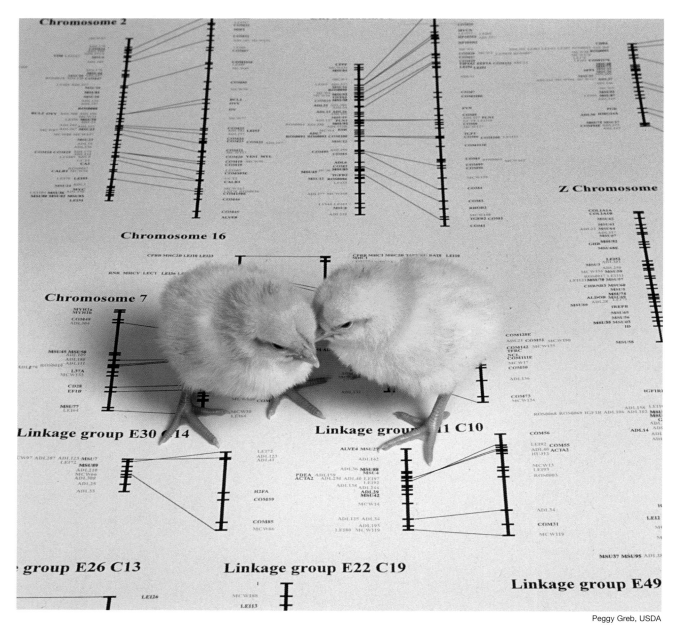

Peggy Greb, USDA

Chicks stand atop a genetic map of a chicken. The new chicken genome will make it easier to locate genes, especially those for complex traits such as disease resistance.

Green jungle fowl, classified as *G. varius* or *G. furcatus* and also called Java or fork-tailed jungle fowl, are considered the most primitive of the four species. They can interbreed with the other jungle fowl and domestic birds but have reduced fertility. Now absent from the island of Java, they inhabit numerous smaller islands and are common in coastal areas and on nearby islets. Green jungle fowl are capable of sustained voluntary flights and observers have documented them moving from island to island. Their diet includes more tidal-zone foods, such as snails and small sea creatures, than that of their inland relatives. They prefer meat foods, such as insects and larvae, to the vegetation preferred by other species. Green jungle fowl that inhabit inland areas forage in rice paddies and other wet environments.

These adaptable fowl are also willing to live in semi-arid environments on small islands. During the dry season, they make do with little access to fresh water. Less terrestrial than the other species, green jungle fowl will spend time in trees. They can make their way along branches and move from tree to tree by jumping from branch to branch.

Cocks have short, truncated necks and hackle feathers instead of the long, pointed feathers of the other three species. They have sixteen tail feathers, compared to the fourteen of other species. Although black overall, the feathers have a blue, green, and bronze metallic sheen and orange-red highlights. The bronze and yellow edges of the neck and hackle feathers make them shimmer green. Hens have penciling on their back and rump feathers, mixing with irregular barring overall. The upper breast feathers have a dark edge over the light lower breast feathers.

Both sexes have a single-blade comb without serrations and a single wattle. The comb is green at its base, shading to purple and red. The male's wattle is red, shading to yellow and purple on the edges; the female's is red and much smaller than the male's. They have pinkish-slate legs and toes.

A biochemical analysis called electrophoresis has made possible refinements of the egg-white proteins of jungle fowl. Using this method, about ten main groups of egg-white protein can be followed to tease out the ancestry of domestic chickens.

Red jungle fowl lack the silver gene common in modern domestic fowl. This dominant gene results in silver plumage, while its recessive paired gene results in gold plumage. Gray jungle fowl have the silver gene, indicating that this species is the likely source for at least that characteristic in modern fowl. Similarly, the extended black gene in green jungle fowl, present in modern fowl, probably came from that species.

Breeders who have crossed various jungle fowl with each other and domestics report varying results, which suggests that the species are more separated genetically than their appearance might suggest.

Red jungle fowl also carry an endogenous retrovirus sequence found in modern fowl but not in other jungle fowl. The retrovirus came with them after they separated into a species, but before they became domesticated. Researchers suggest that this evidence indicates modern fowl descend more directly from red jungle fowl than any of the others.

Charles Darwin, lacking the contributions of biochemistry, asserted that the red jungle fowl was the sole ancestor of modern fowl. Although controversial, his view prevailed for many years.

Mapping of the domestic chicken genome is in progress, but not enough work has been done to establish connections such as these. The technology is there and eventually, as data are collected, answers will become more clear.

CHICKENS AND CULTURE

Chickens, irresistible with their shiny feathers and provocative crowing, have always drawn attention. Their flair made them objects of entertainment and religious significance long before people figured out they were also good to eat.

Fishermen in Indonesia still use jungle fowl to help them navigate the ocean. Known as Ayam Bekisar, these birds are crosses between male green jungle fowl and domestic game hens. They have a long vociferous crow that carries miles across the open ocean. Fishermen stay in touch with each other at night or in the fog by relying on the crowing from their bamboo cages. Island residents hold contests to pick the best crowers.

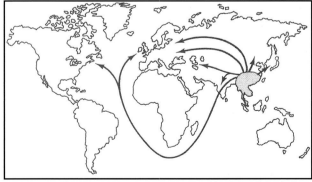

Lewis Stevens, reprinted with the permission of Cambridge University Press

Once domesticated, chickens spread around the world. This map indicates some possible routes along which they traveled, through trade and with human migrations.

Wealthy Chickens

Newspaper king William Randolph Hearst had a touch of Hen Fever, as did many wealthy gentlemen of the era. He added rare poultry to his estate, La Cuesta Encantata, in San Simeon, California. The property is now a state historic monument. Architect Julia Morgan designed an elaborate poultry house for the estate. It has since been refurbished as a house for the ranch manager of the private San Simeon Ranch, still owned by the Hearst family.

People across Asia adopted domestic chickens during the first millennium BC. A rooster graces a coin from Northwest India dated to about 800 BC. From the same time period, chickens appear on seals from Assyria, a region on the upper Tigris River in modern Armenia; the birds played an important role as the guardian of good against evil in the religion of Zoroaster.

From there, chickens show up in Mesopotamia, parts of modern Iraq, Turkey, and Syria, and Asia Minor, the area east of the Bosphorus between the Black Sea and the Mediterranean. They are portrayed on altar stones, still holding more religious significance than economic value.

An Egyptian writer, Thothmes III in the mid-fourteenth century BC, described birds that lay eggs daily, which must have been chickens. As the Persians expanded their empire in the sixth century BC, chickens came with them. The empire eventually encompassed northern Libya to central Asia. The coins of fifth-century BC Himera in Sicily show a cock on one side and a hen on the other.

The Greeks honored chickens and accepted them as a symbol of light and health. Chickens were also used for fighting as sport, which became so popular that for a time matches were organized by the Athenian government. Prolific egg-laying, outside a short breeding season, led to chickens becoming the symbol of fertility. The cock's elaborate courting display became associated with erotic love. Inexpensive and easy to rear, chickens were ideal as food, as well as spiritual nourishment. They could be sacrificed to gods by even the most humble. Aristophanes wrote, "Every Athenian had his hen, even the poorest." Aristotle's *History of Animals* in the fourth century BC shows close observation of chickens. He was sensitive to distinct breeds and their varying characteristics.

The Romans continued to develop the spiritual side of fowl, using chickens to predict the future. A common method was to throw food in front of them and predict success in battle depending on how enthusiastically they ate. Their appetite was easily manipulated by not feeding them for a while before the oraculum was conducted, so the divination method fell into some disrepute. A famous story from the first Punic War tells how the consul P. Claudius Pulcher became angry when the prophetic birds refused to eat before the sea battle of Drepana in 249 BC. He threw them in the ocean, saying, "May they drink if they won't eat."

His subsequent defeat was attributed to his disrespectful act.

Chickens are well documented in Rome. White fowl were required as sacrifice by some gods, such as Anubis, an Egyptian goddess worshipped in Rome, and Zeus. Ovid reports that black fowl were required by Nox, the goddess of the night. Roman experts knew which breeds were the best fighters, which reproduced the best, and which were the best mothers, among other qualities. Columella, Pliny, and Cicero all included chickens in their writings about the Roman world.

The Celts acquired domestic chickens by the first century BC, perhaps from contact with the Romans in northern Italy. Romans introduced chickens to Britain as the Roman Empire expanded. Those birds were Dorkings, the five-toed fowl still raised today. A mosaic from the third century AD from a site on the Danube shows Mercury, the god of commerce, with a five-toed cock.

Marco Polo, in his travels to the Far East, remarked on chickens with feathers like hair in AD 1200—likely the Silkies we know today.

Courtesy of the University of Oklahoma Press

The Italian naturalist Ulisse Aldrovandi wrote the first major work on chickens in 1600. He spent his own money for artists to draw chickens to illustrate his work, such as the Paduan rooster.

This Paduan hen accompanies her mate in Aldrovandi's treatise on chickens. He intended to include everything known about chickens, from botanical terms relating to them to recipes, along with his own observations.

Chickens crossed the Pacific as far as Easter Island and may have made it to South America before Columbus and his chickens arrived in the New World. The distinctive blue and blue-green eggs of Araucanas hark back to their heritage from Collonca and Queteros chickens of South America's Chilean highlands and Peruvian highlands, respectively.

Whether the Ayam Bekisar guided sailors from Oceania to land on islands off South America in the distant past remains an unanswered question. Archaeological evidence of pre-Columbian chickens has not yet been found. Columbus certainly brought chickens with him and returned to Spain with the American turkey, which became popular in Europe. And chickens certainly spread quickly across Central and South America after Columbus.

Darwin's Beagle expedition took him to places in South America where he may have encountered Araucanas, but he did not remark on any unusual egg color. No documentation describes it before the 1920s, so how long it has characterized their eggs isn't clear.

In 1600, Ulisse Aldrovandi, naturalist and professor at the University of Bologna in Italy, wrote a comprehen-

Marco Polo grew up in Venice and traveled in 1271 with his father and uncle to China, where he stayed for seventeen years. The observations he recorded were so fantastic, such as chickens with feathers like hair, that many thought he made them up. The birds he saw must have been Silkies, like this Buff cockerel.

Corallina Breuer

sive treatise on chickens in the second volume of his *Ornithologia*. L. R. Lind of the University of Kansas translated Aldrovandi from the original Latin, giving us access to this encyclopedic work. Aldrovandi spent his life studying various natural subjects, from plants and animals to astronomy and medicine. He was the first to include drawings of the anatomy and skeletons of his subjects,

often spending large amounts of his own money for artists for his books. He remarked on behavior from his own experience, including one hen he raised who lived with him in his house, sleeping near him among his books.

Chickens were an established part of the household in nearly the entire world by then. Easily kept onboard ships, chickens were frequent passengers during the era of

European exploration. They reproduced easily, making them part of every community that settled. American colonists raised chickens, as well as other fowl, and hunted wild birds. Wild fowl and other animals were so plentiful in the New World that early arrivals thought they might have found the Garden of Eden.

By the eighteenth century, distinctive breeds were the norm. Artists included them in bucolic works. They were an important economic part of rural and urban life, contributing to trade as well as diet.

In the United States, the turn of the nineteenth century opened the West to development. The Mormon exodus from Illinois to Utah brought chickens—probably Dominiques, Games, and Dorkings—to the far western settlements.

Individual breeders had long bred their birds for desirable qualities, but the nineteenth century in Western Europe and the United States became the heyday of the Feather Fancy. Asiatic breeds such as the Brahma and the Cochin were imported to the United States in the 1840s. They created a sensation with their dignified carriage and heavy feathering. Javas were imported by midcentury and became a foundation breed for many American breeds. Hen Fever, a humorous name given to the obsession with fancy breeds, became the affliction of wealthy country gentlemen.

By the early twentieth century, chickens occupied urban backyards as well as rural barnyards. They were often the housewife's province, giving her "egg money" and eggs, meat, and chicks to barter. Such economic value played a significant part of the family farm economy. Up until World War I, poultry in cities and suburban areas equaled half the human population.

According to the April 1927 *National Geographic*, which was devoted entirely to chickens, practically every farm in the 1920s had a small flock of fifty to three hundred birds. At that time, poultry was the fifth-largest agricultural commodity, exceeding even beef, wheat, and all fruit products.

Courtesy of the Garfield Farm Museum

The Java was imported to the United States in the nineteenth century and became influential in creating American breeds. Recent interest has reclaimed this breed from nearly disappearing.

First American Poultry Show

The first American poultry show was held November 15 to 16, 1849, at the Public Garden in Boston. It featured popular American breeds such as the Plymouth Rock and the Dominique among the more than one thousand other birds exhibited.

More than ten thousand spectators crowded the show. Daniel Webster was one of the exhibitors, entering a pair of Javas.

Andrew Zimmerman

People have long associated chickens with religion. This Buff Orpington rooster and wheaten hen peck bugs in the garden with a statue of the Buddha.

Lindsay Rowland

Dorkings have five toes, the marker of their breed. They can be identified in Roman art by that fifth toe. This Silver Gray Dorking rooster carries history in his genes.

STANDARD OF PERFECTION

Flock owners, whether effete gentlemen dabbling in a hobby or determined housewives making ends meet, bred their birds to separate them from other flocks. Distinctive birds may have won ribbons for their proud owners at shows, but they were also easily identified if stolen in less elegant circumstances.

Breed standards were formalized upon the formation of the American Poultry Association (APA) and the publication of its first *American Standard of Excellence* in 1874. Now called the *American Standard of Perfection*, these guidelines remain the standard by which chickens are judged at shows.

Breeds have waxed and waned in popularity. Poultry fanciers and enthusiasts develop their own, and the poultry industry pursues birds that will meet commercial needs. Colors and feather patterns vary, resulting in varieties of established breeds. Each serves a purpose in the grand picture of modern poultry.

Through the Depression and up until World War II, small flocks continued to accompany human settlements. Herbert Hoover's "A Chicken in Every Pot" slogan touched the hearts of people for whom that message translated into prosperity.

With the end of the war and the return of the soldiers, American life accelerated technologically. The principles of industry were applied to agriculture and to feeding a hungry nation. Huge broiler and egg houses made chickens industrial units of production. Hybrid Cornish/Rock and other crosses filled the need for units that converted feed into meat and eggs as quickly as possible.

Corallina Breuer

This Buckeye rooster lets fly with a crow. Buckeyes are a quintessentially American breed, developed to withstand cold Ohio winters. Their rich mahogany plumage is reminiscent of the buckeye, from the Ohio state tree.

Courtesy of the American Poultry Association

Courtesy of the American Poultry Association

Poultry artist Arthur O. Schilling painted this Dominique male and female in black and white to illustrate the original American Poultry Association *Standard of Excellence* in 1915. The portraits he painted between 1914 and 1952 are still used to illustrate the Standard.

The American Poultry Association logo includes its founding date of 1873. Devoted to the poultry industry, the APA now functions in exhibitions, setting standards, training and licensing judges, and advocating for "the importance of maintaining purebred stock as a foundation for a progressive and profitable breeding program."

National Poultry Museum

The National Agricultural Center and Hall of Fame in Bonner Springs, Kansas, has a place in its heart for poultry. Its 172-acre site includes a hatchery building in the Farm Town, a 1900-era replica of a farm community that includes a one-room schoolhouse, a general store, a blacksmith shop, and a farmstead, where the center's chickens live.

A new exhibit will include interactive attractions set along a timeline of American agriculture. A timeline helps the public see how agriculture has been entwined with American history and its current importance. Executive director Cathi Hahner finds people are as interested in what's going on in agriculture today as they are in its history.

From taxes on agricultural products during the Revolutionary War through the rebuilding of the South after the Civil War, the exhibit will show how agriculture ties into our history. Technology changes since the 1950s have changed the way agriculture feeds America and benefits other countries through trade.

"People shouldn't think of farmers in overalls with a pitchfork anymore," Hahner said.

Loyl Stromberg, a lifelong poultry lover, has been instrumental in generating interest in a separate National Poultry Museum on the site. His personal collection of poultry memorabilia fills his Minnesota home and garage. He has written many books about poultry. His family continues to operate the family business, Stromberg Hatchery in Pine River, Minnesota.

Stromberg has been a tireless advocate for the preservation of poultry history. Now over ninety years old, he continues to write about poultry and raise funds for the poultry museum.

The center is 18 miles west of Kansas City off Highway 70 and open seven days a week from mid-March through November.

Poultry Words and Phrases

Bad egg: An unsavory person.

Cackle: Derogatory description of loud laughter.

Chicken feed: Insignificant amount of money.

Chicken or hen scratch: Unreadable and ugly handwriting.

Chicken with its head cut off: A lot of activity without direction; acting hysterical or brainless.

Chicken, chicken out, chicken-hearted, chicken-livered: To be cowardly.

Chickens coming home to roost: To experience the consequences of one's behavior.

Cock-and-bull: A fantastic story that is unbelievable.

Cocotte: Prostitute.

Don't count your chickens until they are hatched: Proverb cautioning against spending assets until they are in hand.

Don't put all your eggs in one basket: Proverb cautioning against committing too many assets to a single investment.

Dumb cluck: Stupid oaf.

Egg on your face: To be caught in an embarrassing situation.

Egghead: Intellectual.

Flew the coop: Left the area, usually with disparaging implications.

Go to bed with the chickens, get up with the chickens: Go to bed and rise early, with the sun.

Good egg: A regular guy, good-natured person.

Hard-boiled: Tough, street-smart.

Henpecked: A husband whose wife bosses him around.

Mad as a wet hen: Furious.

Mother hen: A kindly but perhaps smothering woman.

Nest egg: Savings.

Pecking order: Rank, social order from high to low.

Peep: Tiny sound.

Rule the roost: To dominate the group.

Scarce as hen's teeth: So rare as to be unique (hens have no teeth).

Scratch for a living: To scrape and make do, to stretch finances.

Spring chicken: Youngster.

To brood: To be absorbed by negative circumstances.

Walk on eggs: Deal with a situation gingerly to avoid touching off sensitive feelings.

POULTRY IN MODERN LIFE

The industrialization of poultry-raising marginalized back-yard flocks, at the same time suburban life made backyard flocks less viable. By the end of the twentieth century, chickens in the yard were the exception rather than the rule.

Today, commercial poultry operations are selling more than 35 billion pounds of chicken and some 90 billion eggs each year. Approximately 5 billion pounds of chicken is exported, but the rest is consumed here in the United States. The U.S. Department of Agriculture values the broiler industry at $43 billion.

Broilers are raised in all-in/all-out barns, slaughtered at thirteen weeks of age or younger. They are housed in buildings that contain as many as forty thousand birds.

Interest in heritage breeds of livestock of all kinds, including poultry, has increased as the twenty-first century begins. Concern for the origin and conditions of food translates into the next step: growing your own. Even those who recoil from butchering their own chickens feel comfortable eating the eggs of the birds they know and love.

Establishing your own flock puts you in the spectrum of poultry breeders. As others have before you, you will inevitably make your mark on your birds, as they will on you. Whether you eat only their eggs or become a merchant selling meat and eggs at your local farmers' market, you are an important link in a long relationship between humans and chickens.

Courtesy of the National Agriculture Center

America's farming tradition is honored at the National Agricultural Center and Hall of Fame in Kansas. The center's flock of heirloom chickens sets the scene in the Farm Town USA site. Visitors are allowed to help feed the chickens, and the chickens are always willing to cooperate.

Cornish/Rock hybrid crosses like these are used for commercial meat production because they exhibit the fastest conversion of feed into meat. They grow rapidly to feed a nation hungry for chicken.

Corallina Breuer

BREED TYPES

Once you have decided to raise a small flock, you have the exciting decision of which breed or breeds to raise. Your tastes will change as you raise them. Choosing from such varied and beautiful birds may confound you with the range of choices, or you may find yourself lured by a particular breed as soon as you see it. But there are breeds for all needs.

BREEDS FOR ALL NEEDS

Production for meat and for eggs has been bred to a high level. Modern chicken breeds and hybrids offer a wide range of characteristics. They are well documented and easily available from hatcheries and the local feed store. Hybrid crosses, with their associated vigor, can serve well in small-production flocks.

Integrated and sustainable operations may find that traditional breeds offer many advantages in their settings. Traditional breeds have long-standing characteristics that have endeared them to farmers over the years. That very history may make them the ideal birds for your needs.

Modern commercial operations keep chickens for a year or eighteen months at most. Fryers and broilers are processed for the table as early as four weeks, seldom older than sixteen weeks. Longevity is not a factor in commercial all-in/all-out production. For the small-flock owner,

however, it's a good indicator of overall vitality. Longevity is associated with low chick mortality, high productivity, and generally strong constitutions.

BREEDS AND VARIETIES

A breed is a type of chicken, defined by its body conformation, comb, and feather quality. A variety represents a variation of color, comb, muff, tuft, or feather within a breed. Breeds breed true—that is, their offspring are reliably similar to them at least 50 percent of the time. Breeding true is a requirement to be recognized by the American Poultry Association and is included in the *American Standard of Perfection*.

Hybrids by definition do not breed true. They are crosses between two or more separate breeds. If you want to supply your own hybrids, you will need to keep breeding flocks of the parent breeds. Small commercial operations that rely on hybrids may prefer to purchase birds from hatcheries as needed.

The single most important quality that should guide you in choosing which breeds to raise is what you like. You will enjoy most and have the most success with breeds that appeal to you. Chickens have wide appeal, and so many varied breeds are available that you should express your personality in the breeds you raise.

HOW MANY BREEDS?

You may want to start with a single breed. That gives you a chance to become your own expert on them. Many experts, such as longtime breeder Henry Miller, recommend starting with a single breed to gain experience. Few people actually follow that advice, though, including Miller, who kept over one hundred breeds and varieties. His example was more powerful than his words.

Keeping more than one breed can improve your general expertise. It can expose you to the differences between breeds as well as the difference between individuals. If you are keeping pure breeds, rely on the *Standard of Perfection* to learn their characteristics and discern which are desirable.

If you intend to breed your birds, having multiple breeds will require more pens and paperwork. Many breeders change the breeds they pursue over time, finding more satisfaction with one than another. Bruce Lentz, a legendary breeder, approved of keeping multiple breeds, so long as the breeder has sufficient resources to maintain an organized breeding program. Keep as many breeds as you want, but maintain a viable population of those breeds.

HOW TO CHOOSE

Chickens have been categorized as meat birds, egg birds, and dual-purpose birds. If you intend to produce for a market, these are important considerations.

Leghorns are the acknowledged leaders in egg production, but many other breeds, such as Polish, Hamburgs, and Spanish, lay very well. The Braggs Mountain Buff, a modern composite, was developed for its large brown egg. Bovans, a modern hybrid, are often used in organic egg operations.

Barred Plymouth Rocks are becoming the preferred traditional breed for meat production today at Heritage Foods USA, a food broker working with small producers. Cornish have a meaty, stocky build that makes them good for the table. Cornish/Rock hybrids are the dominant commercial chicken. Dominiques and Wyandottes were popular in American history. Dominique-type chickens are documented back to American colonial days, although there is no evidence that they arrived on the *Mayflower*. Dominiques were recognized as a breed by the early nineteenth century. Wyandottes were developed after the Civil War and attracted many followers. The Columbian Wyandotte was admitted to the *Standard* in 1905. Interest continued through the twentieth century, with the Blue variety recognized in 1977. These breeds are again making their name in niche poultry marketing circles.

All chickens lay eggs and are good to eat, so categorizing them is somewhat subjective. Many traditional breeds

Silkies are a true bantam breed, meaning they have no large-fowl counterpart. Their hairlike feathers come in many color varieties, such as this Partridge cock.

Corallina Breuer

Bantams

Bantams are diminutive chickens, usually bred down to 20 to 25 percent of the size of their large-fowl counterparts. Although they are primarily ornamental, many are good egg layers and are valued for that.

Most are small versions of large breeds, but some are distinct breeds or "true" bantams. That includes Japanese, now recognized in nine varieties. The Black-Tailed White was included in the first APA *Standard* in 1874.

Other true bantam breeds are the Belgian Bearded d'Anvers, Belgian Bearded d'Uccle, Booted, Dutch, Rosecomb, Sebright, and Silkie. The Nankin is a true bantam recognized in both single and rose comb by the American Bantam Association.

Bantams have somewhat different formal classifications from large fowl. They are classified by physical characteristics rather than place of origin. This Belgian Bearded Mille Fleur d'Uccle is in the Feather-Legged Bantam class. It has no large-fowl counterpart, so it is considered a true bantam.

Corallina Breuer

provide both and are considered dual-purpose fowl. Games, Dorkings, and Dominiques were the most common breeds on American diversified farms prior to the Civil War.

Size is important to consider. Large fowl can be very large indeed, up to two or two-and-a-half feet tall. They require more space than smaller breeds and bantams.

Bantams are miniature chickens. They have been created in nearly all large-fowl breeds, but some bantams—called true bantams—have no large-fowl correlate.

Nankins, Silkies, Pyncheons, Rosecombs, Sebrights, Dutch, and Belgians such as d'Anvers and d'Uccle are true bantams. The Serama, a recent import, is a true bantam. The Tuzo is a hard-feathered Oriental game true bantam. Jungle fowl are classified as bantams because of their small size.

Plumage varies considerably among breeds. Feathers range from hard and shiny to downy to the hairlike feathers of Silkies. Silkies require more protection from wet

Birds like this Buff Brazilian cock are still fought in South America, where cockfighting, like bull fighting, is legal and popular. Cocks are fought in pits in large arenas.

Horst Schmudde

weather than other chickens because their feathers do not repel water as other chickens do and they can get soaked through and chilled.

Disposition may be the deciding factor. Chickens vary by individuals, but breeds do have general tendencies. Game fowl of any kind will be active and aggressive, even hostile in confinement. They may need separate pens to avoid fights between cocks. With some breeds, such as Aseels, you may need to separate even new hens before introducing them to the flock.

PRODUCTION VALUES

Pressures on breeds may ignore production values like egg-laying and broodiness in favor of feathers and coloration.

Broodiness is often considered undesirable, since broody hens stop laying eggs. However, from a small-flock perspective, it's better to have birds that can hatch and raise their own replacements.

Production values need not be at odds with exhibition and breed conservation. Production has been an integral part of historic breeds. Modern hybrid broilers get bigger faster than traditional breeds, but they are not able to forage for themselves and their plumage is boring. Modern layers generally decline in laying after a year or two, while traditional breeds may continue laying well for years.

"Historically, some Dorking strains could be managed for egg production and give you as many eggs as a Leghorn," said Craig Russell of the SPPA. "While they

39

consume more feed than a Leghorn, they are bigger and can raise you a family of chickens. Differences like that are important and allow breeders to select breeds that fit their needs." In addition to Dorkings, other good dual-purpose breeds include Javas, Delawares, Sussex, and Old English Game large fowl.

One unlikely dual-purpose breed is the Aseel, an Oriental breed rare in the United States. Dr. Charles Everett in South Carolina is enthusiastic about his Aseels. "They are laying machines," he said. "They out-lay anything I have on the place." Aseel carcasses are similar in appearance to modern grocery-store chickens, which can be a marketing advantage.

In eastern Missouri, these rare breeds might well turn up at the farmers' market held on alternating Saturday mornings, April through November. The market caters to small-flock owners who want to buy, sell, or trade; 4-H kids starting projects; and ethnic groups who want live birds, especially dark-skinned birds. "We are never going to draw the industrial market, but we play the hand we have," said Kelly Klober, who has been active in supporting the market since it was started as a 4-H community service project years ago.

FOUNDATION BREEDS

Breeds generally fall into two broad genetic categories: foundation breeds and composite breeds. Composite breeds are those that have been developed from the original foundation breeds. Theoretically, if a composite breed such as the Lamona were to disappear, as it nearly has, a persistent and dedicated breeder could re-create it. If a foundation breed such as the Dorking were to become extinct, however, many of its unique genes would disappear with it.

Marker breeds are those characterized by a specific trait, such as the fifth toe of the Dorking, the hairlike feathers of the Silkie, and the short legs of the Japanese and the Scots Dumpy.

AMERICAN

The American class coalesces around the Java, essentially the type specimen for the class. Not indigenous to America, the Java is an Asiatic breed with some Oriental infusion. It influenced all other composite breeds in the class, giving the class consistency.

Java

The Java came to the United States at least by 1835 from the East Indies (hence the name) and was admitted to the *Standard of Perfection* in 1883. As a high-class market fowl, its breeding contributed to the development of the Black Jersey Giant and the Barred Plymouth Rock. Indirectly, its influence reached many other breeds, including Orpingtons and Australorps. Javas are also probably the source of yellow legs and skin in Dominiques.

Javas had nearly disappeared by the end of the twentieth century, but in recent years, attention from specialty breeders and historical societies has given the breed a second chance. Garfield Farm Museum in La Fox, Illinois, played a significant role in the recovery of the Java breed in the 1990s. Garfield Farm is an 1840s living history farm museum committed to historic stewardship. In the course of its breeding program of Black and Mottled Javas, which are black and white, a pure white strain appeared in 1999. While this variety is no longer recognized by the APA,

Courtesy of the Garfield Farm Museum

Garfield Farm Museum's flock of Black Javas is part of its interpretive mission. Javas were an important breed on nineteenth-century American farms. The White variety, long lost, has emerged from the museum's breeding flock.

Breed Characteristics

Large-fowl breeds are classified by place of origin and judged by this classification at shows. The main classifications are American, Asiatic, English, Mediterranean, and Continental. These classifications reflect the number of birds entered in each and have changed over time. Originally, so many Dorkings were shown that they had their own class.

American breeds are large birds that, apart from Lamonas and Hollands, lay brown eggs. They include Plymouth Rocks, Dominiques, Wyandottes, Javas, Rhode Island Reds, Rhode Island Whites, Buckeyes, Chanteclers, Jersey Giants, Lamonas, New Hampshires, Hollands, and Delawares. Lamonas have nearly disappeared, but occasional reports suggest there may still be some out there. Some breeders are working to recreate this composite breed.

Asiatics include Brahmas, Cochins, and Langshans.

English breeds are Dorkings, RedCaps, Cornish, Orpingtons, Sussex, and Australorps.

Mediterranean breeds are Leghorns, Minorcas, Spanish, Blue Andalusians, Anconas, Sicilian Buttercups, and Catalanas.

Continentals are further divided into fowl of North European, Polish, and French origin. The North European category includes Hamburgs, Campines, Lakenvelders, Barnevelders, and Welsummers. Polish birds include Bearded and Non-bearded Polish. The French category includes Houdans, Faverolles, Crevecouers, and LaFlèche.

Not all breeds fit into any of those classifications, so for judging purposes they are put together into the **All Other Standard Breeds** class. These breeds include Modern Games and Old English Games; Orientals, including Sumatras, Malays, Cubalayas, Phoenix, Yokohamas, Aseels, and Shamos; and Sultans, Frizzles, Naked Necks, Araucanas, and Ameraucanas.

breeders are nurturing it with an eye to campaigning to have it included in the *Standard of Perfection* again.

The farm museum supplies Java eggs to hatch at the Chicago Museum of Science and Industry's Genetics: Decoding Life exhibit. The museum breeds over eight thousand Java chicks each year, supplying thousands of chicks to breeders around the country. Out of those, two brown ones showed up in 2004, remnants of an Auburn variety that disappeared in 1870. The Auburns were significant for their contribution to the Rhode Island Red.

Senior Exhibit Specialist Tim Christakos has shared the rare birds with local breeders who are nurturing the Auburns toward sustainable populations: "We know they are not going to go extinct now."

The Java is a heavy breed, with cocks at an ideal weight of 9.5 pounds and hens at 7.5 pounds. Like many historic breeds, Javas grow more slowly than the industrial hybrid crossbreeds that feed our retail appetite for chicken.

Javas were competitive with Rocks in the past, and selective breeding could return them to their former status. "Except for the remaining commercial strains of Delawares, there is no doubt that many strains of rare American breeds have suffered to some extent by the contraction of the genetic base," said Craig Russell. "However, my efforts with the Java convince me that this is a trend dedicated backyard breeders can quickly reverse, and these breeds should be returned to their former levels of utility and beauty."

Kansas breeder Monte Bowen, who has been influential in cultivating Black and Mottled Javas, recorded a monthly egg-laying average of just under ten eggs. Pullets may start laying at five months of age, early for heavy fowl. However, heritage breeds need to be evaluated by all their traits, not

"What's the rarest breed? What breed most needs attention to be saved?"

Well-meaning people often ask what breed is closest to the brink of disappearance, to know where to focus their energies.

Given the lack of chicken-breed registries, there is no clear answer to that question. Since anyone can and does own chickens, no one keeps track of who has what or how many. Even the USDA's National Animal Identification System (NAIS) won't change that, since it doesn't identify chickens by breed.

The Society for the Preservation of Poultry Antiquities (SPPA) surveys its members and invites all breeders of rare and historic breeds to share their information every year or so for its breeders directory. The directory lists all members who have responded, what breeds and varieties they keep, and how to contact them. It's an invaluable resource for all interested poultry people.

Recent SPPA bulletin editor Ed Hart of Illinois added helpful articles on breeding, buying and selling, and historic aspects. His list of breeds no longer reported serves as a cautionary reminder of the fragility of rare populations.

Identify a breed that has historically been adapted to your geographic location and meets your personal production and aesthetic goals. Those guidelines point to the most likely route to a successful, happy chicken-raising experience.

along a single parameter such as average annual eggs. Bowen resists comparing Java laying ability to other breeds' without acknowledging the Java's overall utilitarian qualities. Javas are good foragers and the hens make excellent brood hens and mothers. They are gentle and patient in disposition.

Because they take more time to develop, Bowen recommends giving them up to a year to mature before culling: "Size must be maintained in this breed, so it would behoove us not to cull too early."

ASIATIC

The Asiatic breeds are a coherent group with a common background. All have heavily feathered feet and legs, even though Cochins and Langshans are quite different.

Cochin

Cochins came to America under the name Shanghai China fowl in 1845. Their size and lush feathers created a sensation in England that became known as the Cochin Craze. Their fluffy feathers make them look even larger than they are.

Originally bred as meat birds in Buff, White, Black, and Partridge varieties, Cochins are now bred mainly for

Corallina Breuer

Blue and Black Cochin bantams are color varieties of the Cochin breed. These soft-feathered birds created a sensation when they were imported to the United States from China in the nineteenth century.

exhibition. The APA *Standard* requires yellow skin and brown eggs. They have since been recognized in Silver-Laced, Golden-Laced, Blue, Brown, and Barred varieties. Fanciers raise Silver Penciled, Mottled, Red, Splash, Birchen, and Frizzle varieties as well.

Langshan

Langshans are an ancient Chinese breed. Smaller than Cochins, Langshans are China's original dual-purpose fowl. They remain the best dual-purpose breed among the Asiatics. Originally only Black, they are now also shown in White and Blue varieties.

Cocks have large tails with sickle feathers as long as seventeen inches. Feathers are close-fitting and smooth. They have white skin and lay dark brown eggs.

ENGLISH

The Dorking characterizes most breeds in the English class. Although not originally British, it came to represent English chickens. The Dorking influence on the rest of the breeds in the class is significant. The Cornish, a composite, is distinctly different and was originally in the Oriental class.

Dorking

Dorkings were known to the ancient Romans, as documented in Roman artwork and literature. The Romans brought them to England in the first century AD, where the Celts and Picts already had chickens but kept them for fighting rather than food. The Roman influence changed that, and chickens became popular livestock.

Their unique marker is the fifth toe, which separates them from other breeds. Silkies also have a fifth toe, suggesting a shared ancestry. In modern European and American composite breeds, the Dorking is the source of the fifth toe.

Dorkings have been important worldwide, with slightly different emphasis in different countries. In America, they were developed along dual-purpose lines.

The APA recognizes Silver-Gray, Colored, Red, White, and Cuckoo varieties, but breeders also fancy Black, Blue-Breasted Red, Brown Red, Buff, Crele, Dark Birchen Gray, Light Gray, Red Pyle, Silver Blue, and Spangled. Historically there were even more old farm colors, such as Brassy Back, Silver Crele, Red Hackles, Silver Hackles, and Cinnamon.

Lindsay Rowland

Dorkings became known as the Old Five-Toed Fowl in England. They take their name from a market town in the south of England.

CONTINENTAL (NORTH EUROPEAN)

Hamburgs, Braekels, Campines, and Lakenvelders are distinct light-laying breeds that originated from the Old Turkish Fowl of the eastern Mediterranean. Hamburgs, Braekels, and Campines reached breed status through efforts of breeders in the Low Countries of Western Europe. Lakenvelders were developed in adjacent Germany.

Hamburg

Hamburgs are Dutch, despite their German name. They were named for the German port of Hamburg, from which

Courtesy of the Claude Moore Colonial Farm, McLean, Virginia

The Silver Spangled Hamburg is a light egg breed popular at exhibitions. Spangled feathers are silver with sharply contrasting black at the end of the feather.

Many colors and both single and rose combs are recognized in Leghorns. This Black rose-comb pullet will grow up to lay lots of eggs, for which the breed is best known.

Jeremy Trost

many of these chickens were coming to England. They were known first there as Pheasant Fowl for their size and lively, active nature. Over time, they came to be known by many names, all of which had their partisans. At the Great Birmingham Show of 1849, exhibitors attempted to resolve the confusion by reaching agreement on a single name. Unable to reach consensus on any of the existing names, Reverend E. S. Dixon suggested Hamburg, a name with no previous association with the breed. It garnered support because while none liked it, at least none of the partisans could claim victory.

Hamburgs are classified with Continental breeds because of their origin, although much of their modern variety development occurred in England. They are known as good egg-layers, although exhibition drove their development. They have white skin and lay white eggs. All have rose combs. Silver Spangled, Golden Spangled, Golden Penciled, Silver Penciled, White, and Black varieties are recognized. Blue and other varieties are also raised by fanciers.

Campine and Braekel

Campines and Braekels share a common history in Belgium. Both are bred in Silver and Golden varieties.

Campines are the white-skinned, white-egg-laying breed in the *American Standard*. Roosters are hen-feathered, which means they don't develop the long sickle feathers, pointed hackle feathers on the neck, or pointed saddle feathers in front of the tail.

Braekels, not recognized by the APA but shown in Europe, are larger and the males have distinct plumage from the females. In the nineteenth century, they were the dominant breed on Flemish farms, popular both as table birds and for their eggs. Its meat is reputed to have a slight game flavor.

The Braekel is a successful forager and resistant to extremes of heat and cold. It became so popular in Belgium that many varieties besides Golden and Silver were developed, eventually standardized, and named for the village, Nederbrakel, that first founded a Society of Braekels. Flocks were severely depleted during World War I, when most were eaten, and the breed never recovered its former popularity. A club, Speciaalclub voor het Brakelhoen, was formed in Belgium in 1971 to save the breed.

Lakenvelder

The Lakenvelder has a black head, black hackles, and a black tail on a white body, with some black on the wings

American Bantam Association Classifications

Bantams are classified in different groupings than large fowl for exhibition showing. Frizzles may be entered in any breed or variety and compete in any class. All other breeds are classified as follows:

Game Bantams

American Game
Modern Game
Old English Game

Single Comb Clean Legged other than Game Bantams

Ancona
Andalusian
Australorp
Campine
Catalana
Delaware
Dorking
Dutch
Holland
Japanese
Java
Jersey Giant
Lakenvelder
Lamona
Leghorn
Minorca
Naked Neck
New Hampshire
Orpington
Phoenix
Plymouth Rock
Rhode Island Red
Spanish
Sussex
Welsummer

Rose Comb Clean Legged Bantams

Ancona
Belgian Bearded d'Anvers
Dominique
Dorking
Hamburg
Leghorn
Minorca
Red Cap
Rhode Island Red
Rhode Island White
Rosecomb
Sebright
Wyandotte

All Other Combs, Clean Legged, Bantams

Ameraucana
Araucana
Buckeye
Chantecler
Cornish
Crevecoeur
Cubalaya
Houdan
La Flèche
Malay
Polish
Shamo
Sicilian Buttercup
Sumatra
Yokohama

Feather Legged Bantams

Belgian Bearded d'Uccle
Booted
Brahma
Cochin
Faverolles
Langshan
Silkie
Sultan

that is not visible when the wings are folded. Lakenvelders have the reverse feathering of Campines, with both sexes taking on what was originally a male feather pattern.

The breed was developed in the Westphalian section of Germany for its egg-laying prowess. It lays white or lightly tinted eggs and is not a good setter. It has white skin. The Lakenvelder's high-contrast plumage makes it an attractive entry in American shows, though rarely seen. The Silver variety is the only one accepted into the *Standard*, but others have been developed by fanciers.

CONTINENTAL (POLISH)
Polish chickens are included in the Continental class by geography, not by type. Their origin is probably Eastern Europe.

Polish

Polish chickens have been known as a breed since the sixteenth century. Their history is undocumented prior to that, but crested chickens are mentioned by the Romans. Their topknot of feathers actually grows from a bony knob on top of the skull. This unique marker identifies them and the breeds to which they are related. Crested breeds also have large nostrils.

Both bearded and nonbearded varieties are recognized. White Crested, Golden, Silver, White, Black Crested White, White Crested Black, and Buff Laced are the recognized nonbearded varieties. Golden, Silver, White, and Buff Laced are the recognized bearded varieties.

Polish chickens have a small V comb. They lay white eggs and today are considered primarily an ornamental breed but were originally an egg breed.

MEDITERRANEAN
Spanish, Minorcas, and Andalusians are recognized as separate breeds but share ancient roots in the Iberian Peninsula. They differ from Italian breeds such as Leghorns in leg color and the genetics of White varieties, although Black predominated in both locations originally.

The English influenced the development of these light Mediterranean egg breeds in the nineteenth century from birds imported from Spain and Portugal. These birds traditionally laid as many eggs as the prolific Hamburg, and their eggs were twice as large. Breeding for showing

Brenda Ernst

This Silver Laced Polish pullet has a large crest of feathers. The origin of the name may relate to the country where it was developed or to the bird's poll, a name for the top of the head in livestock.

rather than production has reduced the remarkable egg-laying abilities of these breeds. They have also lost much of their ability to brood eggs, as this characteristic has been selected against for a long time.

Leghorn

Leghorns originated in Italy, placing them in the Mediterranean class. They are prolific layers and may lay three hundred or more eggs in a year. Broodiness has been bred out of them, so they continue to lay eggs. Most commercial eggs come from Leghorns. Their relatively small size makes them somewhat scrawny as meat birds, but cocks are sometimes raised as fryers.

They have been so popularly bred that many colors are recognized by the APA *Standard*. Originally, they were only White, Black, and Brown. Both single and rose combs are recognized in Dark Brown, Light Brown, White, Buff,

Black, and Silver. Single-comb varieties are also recognized in Red, Black-Tailed Red, Columbian, and Golden Duckwing colors. Fanciers keep other varieties, including Exchequer, a Scottish variety coming into demand at Scottish farms and festivals.

Spanish

Spanish have white faces and white earlobes against striking black feathers. They have dark legs with pink on the bottoms of their feet, compared with the yellow legs

Jeremy Trost

The White-Faced Black Spanish may be the oldest breed in the Mediterranean class. This bantam cockerel has a well-developed white face and an excellent comb.

Jeremy Trost

An attractive White-Faced Black Spanish bantam pullet is a good match for the cockerel. She should weigh about 22 ounces to his 26. Sound breeding maintains the desired characteristics, white face and black feathers, without allowing other qualities to decline.

Stephen Ausmus, USDA

The White Leghorn is the quintessential laying hen. This proud bird is a reliable egg producer for small flocks but is considered too scrawny to be a table bird.

of Italian breeds such as Leghorns. Their white skin and large chalk-white eggs have distinguished them for centuries. The APA considers them the oldest of the Mediterranean class.

Minorca

Minorcas were originally called Red-Faced Black Spanish but were recognized as a separate breed in Black and White Single Comb varieties by the APA in 1888. Rose Comb Blacks and Rose Comb Whites, developed by adding Hamburgs to the breeding line, were admitted later. The Single Comb Buff, the result of adding Buff Leghorns and Brahmas, was also eventually recognized. The arrival of Buff Cochins in the mid-nineteenth century set off a craze that resulted in buff varieties of many breeds. Minorcas are the largest of the three breeds, with size an important qualification. They have white legs.

Andalusian

Andalusians were first imported to England in the mid-nineteenth century. Their blue coloring, a genetic color mixture that never breeds true, excited such interest that Splash and Blue Andalusian cocks were being bred to any black hen even remotely resembling Spanish type to meet the demand. Breeding always produces some Black, some Splash, and some Blue offspring.

Blue chickens were common in England, so Spanish breeds were often crossed with blue games and other breeds. The preferred color is Blue Laced, which features blue feathers with a darker blue edging around the feather. Solid blue and Splash Blue variations are also shown. Andalusians are the smallest of the three breeds.

OTHER STANDARD BREEDS

The Other Standard Breeds classification is, as it sounds, a catchall category. The Old English Games and Oriental classes make some sense being grouped together, but the widely varying characteristics of Naked Necks and Araucanas, formerly in the Miscellaneous Class, make this a difficult class for judges. Finding a formal exhibition location for these very diverse breeds is one of the evolving areas of poultry competition.

Old English Game

Old English Games are traditional birds in Europe and America. They have a long history in cockfighting, which is still practiced around the world. It is illegal nearly everywhere in the developed world but has a strong underground following.

Cockfighting

Chickens have been fought as sport since the earliest domestication. Cockfighting, like other traditional blood sports such as bear-baiting, is rejected in the modern world. However, its appeal runs so deep that the sport continues to attract crowds worldwide.

Cocks are fought either with only their natural spurs or with metal, usually steel, razor spurs attached. Fights with natural spurs tend to be more like boxing matches: serious contests, but both cocks usually leave the pit alive to fight another day.

Not so with razor spurs. They inflict bloody wounds. One cock owner was, with some poetic justice, killed at a match in the Philippines a few years ago by his own cock.

Illegal cockfighters have been implicated in the spread of poultry diseases. The illicit nature of the matches makes it difficult to control.

Gambling is a major attraction of cockfights. Large amounts of money often change hands.

In Southeast Asia and South America, cockfighting is legal. Fighting cocks can have substantial financial value. Since the avian influenza outbreaks in 2004, fighting cocks in Thailand are identified with their own passports certifying their vaccination records.

Corallina Breuer

This bantam Old English Game rooster has the Black Breasted Red color pattern that is traditional for many breeds. Old English Games are popular bantams and excellent specimens are often seen at poultry shows.

They are the foundation breed for Modern Games and have been infused into many composite breeds for their vigor and vitality.

Old English Games are the essence of utility fowl: small for meat birds and average egg-producers, but hardy, vigorous, long-lived, and colorful. Eight varieties are recognized by the APA, but fanciers pursue other varieties that are not recognized, including muffed and tassled versions of all recognized varieties. Recognized varieties are Black Breasted Red, Brown Red, Golden Duckwing, Silver Duckwing, Red Pyle, White, Black, Spangled, Blue Breasted Red, Lemon Blue, Blue Golden Duckwing, Blue Silver Duckwing, Self Blue, and Crele.

Malay

The Malay is a distinctive, tall game chicken, named for its origin in Malaysia. This stern, impressive breed has been included in many modern breeds for its size and strength. The traditional type has a three-curved S-shaped silhouette, evenly divided into neck, shoulder to keel, and legs. It has yellow skin and lays brown eggs. The Black Breasted Red was admitted to the *Standard* in 1883. Five additional varieties have since been recognized: Black, White, Spangled, Red Pyle, and Wheaten females. Fanciers raise other colors, such as Blue Red and Cuckoo.

Shamo

Shamos were developed in Japan from Thai Games, beginning in the seventeenth-century early-Tokugawa period of active Southeast Asian trade. Their meat became popular as "pep food"—considered to confer energy, strength, and vigor—and a favorite of Sumo wrestlers.

After cockfighting was outlawed in Japan in 1924, these birds were bred for exhibition. They have always been docile with people. Along with other important breeds, Shamos are now protected by law in Japan for their historic significance.

Japanese breeders have developed many varieties, but the Shamos shown in the United States reflect their Thai Game heritage. They are closely related to Malays but have shorter legs and are shorter overall. The APA recognizes Black, Black Breasted Red, Dark, and Wheaten female varieties. Fanciers breed several other varieties unrecog-

nized in the United States, including the Ko-Shamo or large Shamo, which traces back to sixteenth-century Japanese traders in Southeast Asia and India.

Sumatra

Sumatras are named for the island of their origin, where they developed as a natural or land race, without much interference from humans. They may be as pure-blooded today as they were when they were imported to America in the nineteenth century. Although an original breed, they have not been used much in breeding modern composites.

The cocks' multiple spurs are desirable in exhibition birds. Their long black feathers shine green. The cocks' flowing tails give them a graceful appearance. They have yellow skin and lay white or lightly tinted eggs.

Phoenix

Phoenix is the name given to Japanese medium- and long-tailed chickens in the United States. Unlike the Japanese Onigadori, which carries a gene that inhibits molting, American Phoenix molt regularly, either annually or every two or three years. Special high perches allow the tail to hang down. Few obtain the twenty-foot tails of their Japanese relatives.

Controversy swirls around the distinction between Onigadori, which are not recognized by the APA *Standard*, and Phoenix, particularly in Germany. Breeders on opposite sides of the Berlin Wall pursued different standards during the years of division. The conflicts have not yet been resolved.

In the United States, the Phoenix is recognized in Silver and Golden varieties in these purely exhibition birds, but fanciers also raise Black Breasted Red, Light Brown, and White varieties.

Aseel

The Aseel (sometimes spelled Asil) is a foundation game breed from India. Its history is so intertwined with the Malay and the Shamo that it's unclear which was the progenitor. The Aseel may date back as far as 3,500 years. An Indian manuscript of that age mentions cockfighting, and this breed was the fighting bird. They are known for their aggressive disposition and willingness to pick a fight. Because of these characteristics, they must be kept in separate pens.

The name Aseel may have come from a word meaning "highborn" in Hindustani, reflecting its valuable, even royal, bloodline, or it may have come from an Arabic word meaning "pure" or "thoroughbred."

Aseels were first admitted to the *Standard*, along with Shamos and Yokohamas, in 1981, as oriental games acquired more followers in the United States. Black Breasted Red, Dark, Spangled, White, and Wheaten female varieties are recognized. Fanciers classify them in three sizes and many local variants.

Naked Neck

Naked Necks carry the dominant gene for featherless necks, with no feather follicles at all on the neck. They also have fewer feathers overall than other chickens. Despite their lack of feathers, they do well in cold weather. The breed was developed in Eastern Europe and perfected in Germany. They are also called Transylvanian Naked Necks, reflecting their Hungarian background. Their origins are rooted in Madagascar Games that were traded north through the Arabian Peninsula. They are sometimes

Corallina Breuer

Naked Necks are also called Turkens. They have less than half the feathers of other breeds, making them easy to pluck. This breed nevertheless stays warm in cold climates and tolerates heat well.

Roosters often coexist happily if the flock has enough room. Overcrowding can cause problems from fighting to pecking and worse. This Light Brahma and Naked Neck have no problem sharing responsibility for the flock.

Corallina Breuer

called Turkens, from the fanciful idea that they are the result of crossing chickens with turkeys.

Red, White, Buff, and Black varieties are recognized. Because the Naked Neck gene is dominant, all offspring of hybrid crosses with this breed will have mostly naked necks, a characteristic called bowtie feathering.

Aracauna

Araucanas originated in South America and entered North American consciousness in 1927, when they were included in a *National Geographic* feature on poultry. The magazine describes them as being "discovered" in 1914. They were later imported to the United States by breeder Ward Bower Jr. of Monroe, New York.

Araucanas lay blue eggs. Khaki-green eggs are the result of the blue gene on a brown egg, indicating a cross with some other breed. It's not a color normally seen but some love it. They are unusual in appearance: rumpless because of a shorter spinal column, with large tufts of feathers on the sides of their heads. They are petite, about half the size of a Cornish, but not considered bantams. A Cornish cock should weigh 10.5 pounds, an Araucana cock 5 pounds, a Cornish hen 8 pounds, an Araucana hen 4 pounds.

Black, Black Red, Golden and Silver Duckwing, and White varieties are recognized. Fanciers raise other colors, including Blue-Breasted Duckwings, in which blue replaces black in the color pattern. Blue feathering has long been associated with Araucanas.

This Ameraucana hen has the full, well-rounded muffs and beard typical of her breed. Ameraucanas were developed in the 1970s and admitted to the *Standard* in 1984. Poultry breeders continue to develop variations on modern and traditional breeds.

Courtesy of My Pet Chicken, LLC

COMPOSITE BREEDS

All other breeds are composite breeds, the result of crossing breeds and selective breeding. To be recognized as a breed, birds must breed true—that is, the progeny must reliably resemble their parents.

AMERICAN

The Plymouth Rock, Dominique, Wyandotte, Rhode Island Red, Jersey Giant, New Hampshire, Buckeye, Chantecler, Delaware, Holland, and, if it is still around, Lamona, are composite breeds judged in the American class. All were created from crossing existing breeds and selecting for populations that breed true.

Plymouth Rock

Plymouth Rocks were developed in Massachusetts after the Civil War and named for one of its most famous landmarks. The breed, in its Barred variety, was recognized in the first APA *American Standard of Excellence*, published in 1874. They were originally shown in the Dorking Class, which has since been subsumed in the American class.

Barred Rocks developed from crossing a single-combed Dominique with a black hen, either a Java or, as there were at that time, a clean-legged Cochin. Barred Rocks and Dominiques have yellow skin, the result of their Cochin heritage. Barred Rocks have single combs.

Plymouth Rocks are useful, active, dual-purpose birds that have attracted many followers over the years. Their eggs range from lightly tinted to dark brown. Their admiring breeders have developed six additional varieties recognized by the *Standard*: White, Buff, Silver Penciled, Partridge, Columbian, and Blue.

Lamonas

The Lamona is an American breed created in the early twentieth century in Beltsville, Maryland, at the U.S. Bureau of Animal Husbandry, which later became the United States Department of Agriculture (USDA). Harry S. Lamon, senior poultryman of the bureau, directed the development of this breed that bears his name.

The Lamona is a composite created by crossing Silver Gray Dorkings, White Plymouth Rocks, and Single Comb White Leghorns. The result is a white bird with red earlobes that lays white eggs. In size, it is larger than a Leghorn but smaller than a Dorking or Plymouth Rock. The large fowl was recognized by the APA in 1933. The bantam was recognized in 1960.

Marion Nash, who served as president of the Society for the Preservation of Poultry Antiquities from 1984 until his death in 1996, worked to save the Lamona. After predators killed thirty of his birds, he turned the survivors—two cocks and four hens—over to breeder Lonnie Miller of Windsor, Missouri. Although formal contact with Miller and his flock has been lost, the line may well have survived somewhere.

Jeremy Trost of Wisconsin started a project in 2000 that redeveloped bantam White Lamonas. In 2005, after nine generations of breeding, he began showing them.

Rumors of existing original flocks of Lamonas persist to this day.

This bantam Lamona was created by breeding the same kinds of birds that were used to develop the original Lamona.

Jeremy Trost

Corallina Breuer

The Barred Rock has a single comb, one of the features that distinguishes it from the Dominique. Barred Rocks are gaining popularity as meat-production birds because the carcass resembles the Cornish/Rock cross sold in grocery stores.

Dominique

Dominiques, or Dominickers, are among the best-known chickens in America. If your grandparents grew up on a farm, they would have known what Dominiques were, if they didn't raise them themselves. The modern Dominique has a rose comb, although single-combed birds were common in the past. It lays brown eggs.

Some claim the Dominique came over on the *Mayflower*, although its name refers to the French island of Dominica. Barred chickens were certainly part of early New England farms. They were probably bred from Scots Gray and Old French Cuckoo chickens, which have white legs and white skin mottled with black—features typical in

Dominiques' early development. Crossing with Javas gave Dominiques their current yellow skin. Dorkings and Hamburgs may have also gone into the mix that resulted in this popular fowl.

Wyandotte

Wyandottes originated in New York State as Silver Laced birds, large fowl with the color and pattern of Sebright bantams. They were originally known as American Sebrights or Sebright Cochins. They have yellow skin and lay eggs that range in color from light to rich brown.

The Dark Brahma/Spangled Hamburg cross that led to this breed gave some of the offspring Hamburg combs and Dark Brahma color markings, which are unacceptable in the breed now. The *Standard* requires all Wyandottes to have rose combs regardless of feather color. They feather quickly as they grow.

Wisconsin breeders developed Golden Laced Wyandottes from a Partridge Cochin/Brown Leghorn cross rooster and a Silver Laced Wyandotte hen. As Wyandottes gained popularity, their advocates developed White, Buff, Black, Partridge, Silver Penciled, Columbian, and Blue varieties, which are now recognized in the Standard. Fanciers raise Cuckoo, Barred, Buff Laced,

Corallina Breuer

Silver Lacing is the black stripe around the edge of silvery-white feathers. Silver-Laced Wyandottes like this hen were one of the breeds recognized for exhibition in the late nineteenth century.

55

Violet Laced, Red, Blue Laced Red, Buff Columbian, and other unrecognized varieties.

The Columbian color pattern that now graces varieties of many breeds got its name from the Wyandottes exhibited at the 1893 Columbian Exposition of the Chicago World's Fair.

Rhode Island Red and Rhode Island White

These two varieties share a type but are exhibited as separate breeds. Developed in the early twentieth century from different backgrounds, they are named for the state where they were developed. The APA pinpoints their origin to the section between Narragansett Bay and Buzzard's Bay.

The Red is the result of crossing Black Breasted Red Pheasant Malay, Leghorn, and Asiatic breeds. The original White came from Partridge Cochins, White Wyandottes, and Rose Comb White Leghorns. Some modern White strains have been selected from sex-linked crosses that involved Reds.

The horizontal, oblong body type of these dual-purpose breeds is the same for both. Reds may have single combs or rose combs; in the White breed, rose combs are standard, though single combs still exist. Both Reds and Whites have yellow skin and brown to dark brown eggs. Feathers are smooth and firm.

Jersey Giant

Javas were crossed with Dark Brahmas and Black Langshans to get a meat breed in the late nineteenth century. In this century, Cornish influence has been bred into some strains, adding to Jersey Giants' muscular bulk. The goal was increased size, but the resulting Giants are not proportionately as meaty as Javas. They have always been good layers and have developed as a dual-purpose breed. Like the Java, its skin is yellow and its eggs medium to dark brown. A White variety emerged in the twentieth century. Both varieties are solid color. Blue is now also recognized.

New Hampshire

New Hampshires were developed by breeders in that state who consistently bred their Rhode Island Reds to the goals of early maturity, large brown eggs, quick feathering, strength, and vigor. Selective breeding from around 1915

resulted in birds that were distinct from the original breed and admitted to the *Standard* in 1935. They satisfy both meat- and egg-production goals. Some strains are bred more for meat production and do not lay as well as the dual-purpose strains. Strains selected for egg production are usually smaller than the standard hen weight of 6.5 pounds.

Buckeye

Buckeyes were developed in Ohio, the only North American breed developed by a woman, Nettie Metcalf of Warren. She started developing a distinctive Buckeye State breed from the Rhode Island Red in 1896. She crossed Barred Plymouth Rocks with Buff Cochins, added some Black Breasted Red Game, and then bred the Reds to each other until the color and type stabilized. The existing strains of pea-combed Rhode Island Reds were absorbed

Corallina Breuer

The Buckeye is the only breed credited to development by a woman. It has a small comb to withstand cold winters in the Buckeye State.

into this breed. The breed was admitted to the *Standard* in 1904. "They are big enough to produce some meat, but pretty good layers," said Craig Russell. "It's a good dual-purpose breed, more than simply the meat bird that she tried to create."

Buckeyes' small pea combs withstand the cold well, compared to the single combs found on most American breeds. Frozen combs don't regenerate and the experience is stressful for the chicken.

The birds are a rich mahogany bay, slightly darker on the wing bows in cocks. Think of the rich color of buckeye nuts. Some black marking may occur on cocks' flight and main tail feathers. The undercolor is red, except for a bar of slate across each feather below the surface on the back. The bar is one of the characteristics that distinguish this breed from Rhode Island Reds.

Cocks should weigh about 9 pounds, hens about 6.5 pounds. They have stout muscular thighs when selected for meat production and a broad, well-rounded breast carried well up. The Buckeye should retain the stocky Indian Game–type body that makes it a distinctive breed and a good meat bird. Their game background accounts for their assertive nature and makes them confident around people. They are adaptable to any living circumstances but prefer room to roam.

Holland

Hollands are based on stock originally brought from Holland. White Leghorns, Rhode Island Reds, New Hampshires, and Lamonas were added to the original stock in the 1930s by Rutgers Breeding Farm.

Barred Hollands were developed from White Leghorns, Barred Plymouth Rocks, Australorps, and Brown or Black Leghorns. They were admitted to the *Standard* in 1949. The Barred variety proved to be the most popular among farmers. The White is probably extinct. California Grays are a variation on the Barred Holland created from Barred Rocks and White Leghorns by another company selecting for the gray type. Gray males are lighter than the ideal Barred Holland.

The Holland is a heavy breed that lays white eggs. The type is not the same but they are essentially the same size as a Rock. "They were developed so the small farmer who didn't have a good market for brown eggs would have a white egg from a meaty bird," said Craig Russell.

"They are a good bird for homesteaders and small acreages," said Duane Urch, a master breeder and founding member of SPPA in Minnesota. "They run and are good foragers. They're not a timid bird, but they're not aggressive, either."

Delaware

The Delaware is a result of breeding the almost-white individuals, sports, that occasionally result from breeding Barred Plymouth Rock cocks on New Hampshire hens. Delawares have a unique pattern of black barring in the hackle, primaries, secondaries, and tail feathers.

"They are a striking breed," said Erin Traverse, who, with his wife, Patricia, keeps them at his Vermont farm. "The Barred Columbian pattern makes them more interesting, whether free-ranging across a carpet of green grass or in the showroom, than simply a white chicken."

The Delaware was bred in the Delmarva Peninsula for the production of broilers. They were an important part of the broiler industry in the 1940s, '50s, and even into the '60s. Small farmers then often kept some Delawares to make their own crosses. The breed was admitted to the *Standard* in 1952. They are used in producing sex-linked offspring, by mating Delaware hens to New Hampshire or Rhode Island Red cocks. The day-old chicks can be

Sports

Sports are the unusually colored offspring that occasionally appear from the mating of breeds that typically breed true. Their appearance reveals the existence of otherwise unexpressed genes. Sports are often significant in breeding new varieties or lines of a recognized breed.

Corallina Breuer

The Delaware was developed as a broiler breed from Barred Rocks and New Hampshires in the 1940s. The Columbian barred pattern that occasionally resulted is so attractive that it achieved its own exhibition status; however, the genetics tend to produce hens without the required black in the tail.

Corallina Breuer

The Light Brahma pattern is the same as the Columbian color pattern on American breeds.

separated by distinctive light-colored males and red females. The light-colored males are raised as broilers.

Jord Wilson of Prairie Grass Poultry in Oklahoma started his flock of Delawares four years ago. "It was a rare American breed I wanted to help out," he said. He finds them excellent layers of large dark brown eggs, with light skin that is easy to pluck. They lay from Thanksgiving until after the heat of summer, when they take a break for molting. They grow fast and taste great. Traverse observes that the large to jumbo eggs often have double yolks, which will not hatch. That liability may be overcome if there is a market for such large eggs.

"They are real personable," Wilson said. "As chicks they come right up to you. They are curious but gentle and not flighty as adults. They are one of my favorite breeds."

Delawares are also hardy. When coccidiosis infected part of Wilson's flock, he lost all the Houdans but none of the Delawares.

Chantecler

Developed as a distinctive Canadian breed, this composite thrives in cold weather. Brother Wilfred Chatelain held a doctorate in agronomy when he began working to develop the Chantecler in 1907. The Cistercian Abbey of Notre Dame du Lac's flocks comprised about a dozen breeds. He envisioned a practical, white, dual-purpose breed with a small comb to withstand Canadian winters. A breakthrough in 1917 brought together a 7.75-pound pullet who laid ninety-one eggs in four months and a 10-pound White Plymouth Rock cock. The breed was admitted to the *American Standard* in 1921.

"Here, where 30 degrees below zero is common for days, even weeks at a time, frozen combs on Chanteclers are unheard of," said Traverse, who keeps Buff Chanteclers in Vermont.

The Chantecler's demise was heavily publicized in 1979. The University of Saskatchewan claimed that the last

Getting Recognized in the *Standard*

The American Poultry Association (APA) has a detailed process for recognizing a breed. Marans are currently under consideration.

Birds of the breed applying for recognition must be shown at APA shows at least twice each year for two years. At least two hens, two pullets, two cocks, and two cockerels must be shown.

Judges then submit their opinions of the breed and a qualifying meet is held. No fewer than fifty birds must be shown at the meet. Judges expect the birds to resemble each other closely to establish the breed type.

Varieties such as Cuckoo tend to be less similar to each other than solid colors such as White and Black.

Chantecler rooster, which lived in the Department of Animal and Poultry Science there, had died. While there were no Chanteclers at universities or commercial hatcheries, small-flock owners still had them.

White and Partridge are the recognized varieties. Buffs are raised by fanciers. A Black variety, originally called Albertans for the Alberta Agricultural Station where they were developed, was denied separate breed status; instead, they were admitted to the *Standard* as a variety of Chantecler. This variety is also thought to have died out, but Canadian fanciers are working to re-create it. Chantecler breeders are working on other varieties, and Ideal Poultry now lists a Red variety.

"You get some very interesting colors when you breed a White Chantecler male to a Partridge Chantecler hen," said Traverse.

They are alert and active, although they adapt well to confinement, too. "It was [Wilfred's] desire to give his country a breed of poultry with personality, character, a particular quality," writes Linda Gryner in her 1996 book, *Chantecler*

& Other Rare Poultry Breeds. "If an animal comes into the yard, or a vehicle into the driveway, the males will crow to tell us," said Traverse.

The standard weight of Chanteclers is 8.5 pounds for cocks, 7.5 pounds for hens. Traverse, with twenty years of experience as a chef, finds the meat as delectable as the finest of Indian Games, Old English Games, Dorkings, and Houdans. Like Delawares, they can be butchered at any age.

The Traverses' hens average 180 to 200 eggs a year. "Up along the Canadian border and points north, the laying ability of this breed is very much appreciated," he said. "Patricia and I keep Chanteclers because of their history. Their location makes them a regional heirloom. They are endangered. They are an economical, dual-purpose breed, and they are calm, gentle, personable birds."

ASIATIC

Cochins and Langshans are considered foundation breeds. They have been bred into many color varieties over the years. Brahmas are the only Asiatic composite breed recognized by the *Standard*.

Brahma

The name Brahma emerged from earlier names—Chittagong, Gray Shanghai, and Brahma Pootra—given to

these birds after they arrived in America in the 1840s from Shanghai, China, and the Chittagong area of what is now Bangladesh. They are believed to be originally a Shamo or Malay/Cochin cross but were considerably modified by New England breeders before the Light and Dark varieties were recognized by the APA. The Buff variety is a twentieth-century creation.

Corallina Breuer

This Brahma pullet has Partridge feathers, an unrecognized variety. Many fanciers enjoy varieties that are not recognized by the *Standard* but please their eye. Unrecognized varieties can be exhibited but cannot win awards above its breed or class.

Corallina Breuer

This Dark Brahma cock shows the breed's impressive and imposing carriage. The majestic Brahma is good at foraging to find its own food.

They have yellow skin and lay light brown to dark brown eggs. Their meaty frames make them desirable meat birds as well. Their feathers are smoother and not as soft as their Cochin ancestors.

ENGLISH

The Dorking characterizes most breeds in the English Class. Although not originally British, it came to represent English chickens. Its influence on the rest of the breeds in the class is significant. The Cornish, a composite, is distinctly different and was originally in the Oriental class.

RedCap

Looking at a RedCap tells you where this breed got its name: the large rose comb. They probably were developed from Hamburgs, Dorkings, and Games and trace back to Derbyshire and Yorkshire. They have white skin and lay white eggs.

Cornish

The original Dark Cornish is named for Cornwall, where it emerged from breedings of Aseel, Black Red Old English, and Malay. Its distinctive stocky body, the same for cocks and hens, make it a leading meat bird. Cornish breeding has influenced the modern commercial industry, which overwhelmingly raises Cornish/Rock crosses to supply the market.

Its hard, narrow feathers give it brilliant colors. The APA recognizes Dark, White, White Laced Red, and Buff varieties.

Orpington

This breed is named for the English town of Orpington in the county of Kent, where it was developed late in the nineteenth century. Black Langshans, Black Minorcas, and Black Plymouth Rocks went into the first Black Orpington. Later crosses with other breeds produced White and Buff varieties. Black Orpingtons crossed with Splash varieties of other breeds produced a Blue variety.

They are large, heavily feathered birds suited to meat and egg production. They have white skin and lay light to dark brown eggs.

Australorp

This Black Orpington variation was developed Down Under for egg production. It remains a meaty medium-weight bird, more like the original Orpingtons that were imported to Australia than the modern birds of that name. It has white skin and lays tinted eggs. Its legs are black with pink on the undersides of the feet.

Sussex

The modern Sussex was developed in the English county that gives the bird its name around the turn of the twentieth century. As a meat bird, this breed remains influential in commercial breeding, as well as being popular as a table fowl in Britain and Europe. In the United States, it is popular as a small-flock and show breed. It is the result of Dorking, Game, and Asiatic crossbreeding.

Speckled, Red, and Light varieties are recognized in the United States. Additional varieties are recognized in England.

CONTINENTALS (NORTH EUROPEAN)

Barnevelder

Barnevelders were developed to lay large dark brown eggs, which they do very well. Attention was given to standardizing their markings later, and they were recognized by the *American Standard* in 1991.

They were developed in the Barneveld district in Holland. They remain very popular in Holland and have followers in the United States. The ideal egg color can be influenced by breeding to be a very dark chocolate brown.

CONTINENTALS (FRENCH)

Faverolles

Faverolles are a French breed developed from the Dorking, as evidenced by its five toes, and Asiatic stock, either Brahma or Cochin. The muff is often attributed to the Houdan, but some strains of Dorkings were muffed when Faverolles were developed.

They lay light brown eggs and are known for laying through cold weather but are large-bodied and make a good meat breed.

Faverolles are the only APA-recognized breed with the Salmon color variety, a Silver Wheaten pattern, which differs on cocks and hens. The male looks like a dark

Corallina Breuer

These Salmon Faverolles pullets are pretty and demure compared to the roosters of their breed. This French breed is the only one with feathers this delicate salmon color, which ranges from light pink to copper.

Corallina Breuer

The Salmon Faverolles rooster stays protectively close to two pullets. His dramatic coloring differs from the pale females.

Dorking and females are similar to Wheaten game hens. White is also a recognized variety in the United States, and other colors are recognized in France.

Houdan

Houdans also display their Dorking background in their five toes. Crevecoeurs and Polish probably contributed the comb and crest.

The Houdan has white skin and lays white eggs. It is considered a dual-purpose breed, highly thought of by French cooks and gourmets.

Both Mottled and White varieties are recognized.

Crevecoeur

Crevecoeurs are a crested breed, taking their name from a village in Normandy where they are believed to have been developed. They probably developed from the same background as Houdans but lack the fifth toe. Black is the only APA-recognized color, but fanciers also raise Blue and White varieties. With white skin and producing white eggs, they are a dual-purpose breed similar to Houdans.

La Flèche

La Flèche were bred from Black Spanish, Crevecoeurs, and DuMans in the La Sarthe Valley in France as a dual-purpose breed. The rich pastures of the valley are credited with developing this breed's white flesh.

American and British strains have been developed mainly for egg production. They all lay white eggs.

Most are Black with the white earlobes of Black Spanish, but unrecognized White and Blue varieties are raised by fanciers.

Welsummer

Welsummers were developed from Partridge Cochins, Partridge Wyandottes, and Partridge Leghorns, with some Barnevelder and Rhode Island Red influence. The breed is named after the village of Welsum in the Netherlands. Leghorns were added to result in an excellent layer of large brown eggs. Its eggs are the dark brown of the Barnevelder.

Like other egg breeds, it rarely gets broody. Its size of 6 pounds for hens make it a good dual-purpose breed.

It has yellow skin and a single comb.

Corallina Breuer

Welsummers are a Dutch breed named for a town in the Netherlands. It is known for its large dark brown eggs and is considered a light breed, although roosters such as this one may weigh six pounds.

MEDITERRANEAN

Ancona

Anconas share the background of Leghorns. In Europe, both breeds are known as Italian. They take their name from the Italian city from which they were imported to England in the mid-nineteenth century. Like Leghorns, they are excellent egg-layers with little broody instinct.

They have yellow skin and lay white eggs. Single and Rose Comb varieties with black and white mottled feathers are recognized by the APA. Blue, Brown, and Red Mottled varieties are raised by fanciers.

Sicilian Buttercup

Sicilian Buttercups take their name from their cup-shaped combs but were actually developed in North Africa.

The English imported them from Sicily and identified them with that island. They were imported to America in 1835 but did not receive much attention until the end of the century. They were admitted to the *Standard* in 1918.

Their golden feathers are spangled with black, giving the hen in particular a striking appearance. Roosters have a black tail and some black in the cape. Barry Koffler of FeatherSite considers the hens the prettiest birds in his collection. Other colors are raised in North Africa and Europe.

Catalana

Catalanas include Cochins in their mixed heritage, giving them a hefty build for an egg breed. Originally developed near Barcelona, they are popular in Latin America for both meat and eggs. The only variety recognized in the United States is a black-tailed Buff, but a black-tailed White variety exists in Europe. They have pinkish white skin and lay white or lightly tinted eggs.

Corallina Breuer

This Catalana hen lays plenty of large white eggs. The dual-purpose breed is popular in South America but is acquiring a following in North America.

The full name of the Catalana is Catalana Del Prat Leonada, named for its origin in the Prat region near Barcelona in Catalana, Spain. It was first exhibited at the Madrid World's Fair in 1902.

Corallina Breuer

OTHER STANDARD BREEDS

This class accommodates breeds that do not fit into other existing classes. The Modern Games are distinct but reflect their heritage to heirloom games. Cubalayas relate to their Oriental Game relatives. Frizzles have unique feathering. Ameraucanas are a modern composite derived from the Araucana, with its blue eggs. These birds are combined into a single class because none generally attracts enough entries at a show to be a meaningful class.

Sultan

Sultans originated in Turkey, arriving in England in 1854 from Istanbul as "Sultan's Fowl." They are considered an exhibition fowl, raised primarily for their showy all-white appearance, with full crests, muffs, beards, and feathered hocks, probably the result of a Polish/Silkie cross. Production qualities have never been important, although the hens were originally good layers of medium-sized eggs.

Modern Game

Modern Games were developed as an exhibition breed in England after cockfighting was outlawed in 1835. Their graceful appearance suggests modern sculpture. That carriage is important in the show arena. Nine varieties are recognized: Black Breasted Red, Brown Red, Golden Duckwing, Silver Duckwing, Birchen, Red Pyle, White, Black, and Wheaten.

Cubalaya

Cubalayas are Cuba's contribution to game fowl, based on Oriental birds probably from the Philippines. The traditional game variety Black Breasted Red was the first recognized in 1939. Black and White varieties are recognized and breeders raise other unrecognized varieties such as Light Gray and Red Pyle, as well as Silver, Golden, Blue, Blue Breasted, and Red Duckwing variations.

The white meat of these small birds is considered a delicacy. Their distinctive low, drooping tail is called a lobster tail. Lower tail feathers fan out to overlap the ones above them, forming what Oriental Game expert Horst Schmudde describes in his book on the subject as "an elegant cape."

Yokohama

The APA *Standard* type of Yokohama is a long-tailed game fowl descended from Japanese Minohikis developed in Germany. The Red Shoulder variety has red shoulders over a white saddle. The breast is red speckled with white. A White variety is also recognized.

The APA requires Yokohamas to have cushion combs, but in Japan, pea combs are also acceptable. As with other game fowl, many varieties, colors, and both combs are bred by fanciers.

Corallina Breuer

The Red-Shouldered Yokohama is one of several long- and medium-tailed breeds developed in Japan. An all-white variety is also exhibited. Their long tail feathers require special care.

Ameraucana

Ameraucanas are related to Araucanas but have been bred to have a rump and tail. They have muffs below and around the eyes but no tufts at the ears. They also lay blue eggs.

They were developed in the 1970s as a dual-purpose production breed. They now have their own breed recognition in eight color varieties: Black, Blue, Blue Wheaten, Brown Red, Buff, Silver, Wheaten, and White.

Corallina Breuer

These standard Blue Ameraucanas were developed from Araucanas but have a beard and muffs rather than ear tufts. This dual-purpose breed has a meaty body and lays those popular blue eggs.

Araucana hens like this one have tufts of feathers protruding on both sides of the neck.

Corallina Breuer

UNRECOGNIZED BREEDS

Other breeds not recognized by the APA *Standard* but popular with small-flock owners:

Fayoumi chickens are a small breed raised along the Nile in Egypt. They have speckled feathers and upright tail carriage. They tolerate heat well and are good layers of small, tinted eggs.

Madagascar Games actually originated in Southeast Asia but first came to European notice in Madagascar, hence the name. They are a large Malay-type fowl. They share the sparse feathers of the other Oriental Games and have naked necks. Color breeding is not standardized, but most are Black, Brown Red, Black Breasted Red, Blue, Blue Red, and Orange Red.

Ga Noi is a Vietnamese game fowl in naked-neck and feathered-neck varieties. It is a recent addition to the American scene, reportedly from eggs brought to this country in 1990. It is a fierce fighter with strong bones. Smaller than the Madagascar Game, it is a typical Oriental of the upright type.

Iowa Blues were developed in Iowa around the turn of the twentieth century. Until World War II, developing local composite breeds was common, and the Iowa Blue is a surviving example. Campines and Plymouth Rocks probably went into its development.

Jaerhons were standardized in the 1920s in Norway from native chickens. They are small birds but lay large white eggs. Hens have barred and spotted feathers and cocks are barred on the breast and legs over white on the rest of the body. Dark Brown and Light Yellow varieties are raised by fanciers of this hardy breed.

Marans are recognized in their native France. They can thrive in wet climates, having been developed around the marshy French town of Marans. Their dark chocolate-brown eggs make them attractive to fanciers. The *French Standard* recognizes eight color varieties and is considering recognizing a ninth. Cuckoo is the original variety and still the most common. French Marans have lightly feathered legs but English Marans are clean-legged.

Kraienkoppes are the result of crossing Leghorns and the Pheasant Malays from the Dutch colonies in Indonesia. The breed was perfected as a show bird in the mid-nineteenth century in Holland and Germany after cockfighting was outlawed. They have small walnut combs on the strong head of the Malay. Although many colors were raised in the past, today fanciers mainly raise two varieties: Silver and Black Breasted Red.

The **Orloff** is a hardy Russian game breed that reflects its Malay background in its upright stance. The APA once recognized the Black variety as Russians, but it was dropped due to lack of interest in the breed around the turn of the twentieth century. Bantams are recognized in Red or Mahogany, Spangled, and White by the American Bantam Association (ABA). Orloffs were originally developed as a heavy meat breed, but American and German breeding has modified them to improve egg production and they are now considered a dual-purpose breed.

The *British Standard* recognizes four varieties: Black, Mahogany, Spangled, and White. Black Breasted Red, Mottled and Buff, Cuckoo, and Crele are raised by fanciers. The Mahogany Orloff is described by nineteenth-century writer Edward Brown as having the richest feather color.

The **Tomaru** cock is known for his long and musical crow. The record is twenty-five seconds. Most crow ten to fifteen seconds, with twenty seconds not unusual. It has a two-toned call, ending with a *schnork* sound. These black

Corallina Breuer

Feather patterns such as Cuckoo, exhibited on this Marans rooster, are more difficult to breed true than solid colors. Cuckoo is the traditional variety for this French breed. It is more popular in its home country, where chefs consider it the tastiest fowl.

birds are the largest of the long- and medium-tailed breeds, but their tails average only three to four feet. Their long legs and erect carriage give Tomarus an elegant, statuesque appearance.

Scots or **Scotch Dumpys** take their name from their short legs, only two inches long. Despite their size, they are good layers and pack plenty of meat on their short frames. Their British roots suggest a relationship to Dorkings, which they resemble in type, apart from the legs. Like the fifth toe in Dorkings and the Polish crest, they are a marker for their unique characteristic. Eggs range from chalk white to light cream to light brown. Although Cuckoo is the traditional color variety, Scots are bred in many other colors in their native land. In North America, most fanciers raise Cuckoo, Black, and Dark Gray.

Corallina Breuer

This Cuckoo Marans hen lays dark chocolate-brown eggs, one of the unusual qualities that make its fanciers love the breed. The color makes it easy to pick the shells off boiled eggs!

CHAPTER 4

• •

OBTAINING STOCK

Once you have decided what breeds you want to work with, you can start looking for stock. You can start with hatching eggs, day-old chicks, young pullets and cockerels, or adult birds.

Starting with eggs or chicks allows the birds to establish their social group from the beginning. Pecking order is important to chickens. Adding birds later tends to upset the social order and has to be handled carefully.

On the other hand, acquiring older birds, young pullets, and cockerels allows you to avoid getting birds of the sex you don't want, usually roosters. It also gives you some idea of what the mature bird will be like. You may be looking for specific characteristics to add to your breeding program. You know exactly what you are getting in an adult bird. All new stock should be quarantined for two to four weeks before being introduced to the flock to avoid introducing diseases along with them. After this period, the new birds can be introduced to the flock, separated by chicken wire or other fencing before putting them directly in the pen or yard.

GETTING STARTED

Internet resources are a good place to start for the basics. See the appendix for a full listing of useful websites.

Conservation organizations such as the American Livestock Breeds Conservancy (ALBC) and the Society for the Preservation of Poultry Antiquities (SPPA) provide direct ways to contact breeders. ALBC includes contact information for the breeders they identified in the census. The SPPA publishes a breeders' directory that lists stock owned by all its members.

The rarest breeds may be impossible for a novice to acquire. Breeders will want to know your background and qualifications before they share their stock. The good news is that once you establish yourself as a serious breeder, they will be happy to share stock and advice with you.

Courtesy of the Society for the Preservation of Poultry Antiquities

Many rare and historic poultry breeders belong to the Society for the Preservation of Poultry Antiquities. SPPA holds regional meets at poultry shows, where members have the opportunity to be recognized for their accomplishments. The breeders directory lists members and what they are breeding with contact information. It's an invaluable resource for small-flock owners.

Baby chicks are easiest to ship because they need no food or water for the first 72 hours after hatching. They are usually shipped in batches of 25 or more to maintain warmth.

Joe Valbuena, USDA

HATCHERIES

The business of rare poultry attracts those who are sufficiently devoted to their birds to attempt to overcome the difficulties. Dealing in live animals of any kind is challenging, but recent problems with threats of disease, animal rights protests, and reluctant shippers have increased the difficulties.

Hatcheries, nevertheless, continue to conduct business, offering a direct route for fanciers both novice and experienced to obtain stock.

Some, such as Murray McMurray Hatchery in Iowa and Ideal Poultry Breeding Farms in Texas, have been in business for many years.

Hatcheries have the advantage of a business structure to provide support. They have efficient means of packaging and shipping their eggs and birds, creating reliably satisfied customers. They have knowledgeable people to answer telephone calls and e-mails.

Start with their attractive catalogs. Like seed catalogs, they are springtime dream books. Consult poultry publications like the *Poultry Press* for advertisements.

Day-old chicks are the easiest to ship, making them the preferred stock. Chicks continue to absorb the yolk for 36 to 72 hours after hatching, so they need no food or water.

Their main requirement is warmth. Day-old chicks are generally not available during cold months and are shipped in lots of twenty-five or more to keep each other warm. If this number is more than you plan to order, you may be able to put an order together with other fanciers.

LOCAL RESOURCES

Chickens always do best in the climate in which they were hatched and raised. It's to your advantage to find local sources. The county extension agent, listed in the telephone book under Agricultural Services in the county listings, may be able to refer you to these sources.

The extension agent also oversees the 4-H program. 4-H leaders may advise members who are raising poultry and be able to identify sources of stock for you.

The folks at the feed store may know local sources. They sell flock-keepers poultry feed and are often involved in local agriculture activities.

In rural areas, high schools have agriculture classes, FFA, and 4-H programs. The agriculture teacher or advisor may be able to refer you to local poultry sources.

If you are looking for a rare or historic breed, formal resources within the commercial system may not know what you are talking about. They are likely to have connections to hybrid crosses and offer strategies of industrial farming more so than helping with the nuances of small-flock husbandry.

Poultry organizations exist at all levels. As you become more involved with poultry, you will want to join one or more.

National poultry and livestock organizations such as the **National FFA Organization, 4-H**, the **American Poultry Association (APA)**, and the **American Bantam Association (ABA)** help administer poultry activities. The **Society for the Preservation of Poultry Antiquities (SPPA)** concerns itself solely with domestic poultry. The **American Livestock Breeds Conservancy** addresses all livestock breed conservation issues.

Breed clubs are often international in scope. They focus on information about a particular breed or group of breeds.

They may maintain contact with similar breed organizations in other countries or have international membership. Local chapters may be active in your area. The SPPA has members in countries beyond the United States and has contacts around the world.

The advent of the Internet has improved international communications. Individuals can connect with each other through special-interest sites.

Organizations publish newsletters and directories of members. They can be invaluable in locating rare stock and forging relationships based on mutual interests.

Ken Hammond, USDA

Handling chicks from the start helps them become accustomed to humans. They will be better show birds and companion animals.

SPECIALTY BREED ORGANIZATIONS

Every breed has its advocates, and they usually form breed clubs that provide regular newsletters about their breed and sponsor meets at poultry shows.

Breed clubs advertise in poultry publications such as the *Poultry Press*. They also publicize their shows and meets through the *Press* and other publications. Contact information is available in advertising and regular columns, as well as the appendix of this book.

Joining a specialty-breed organization puts you in contact with the breeders most active in promoting their breed. Having access to them is invaluable. Poultry people are accommodating about answering questions and helping their fellow breeders get started and solve problems along the way.

SHOWS

Attending a show can also put you in touch with local breeders who are raising the breeds you are looking for. Knowing the breeder gives you the advantage of having someone to call for advice later on.

Poultry exhibits at state and county fairs give you an opportunity to see rare and historic breeds. Their owners will be close by, so you can talk with them. Poultry people are usually gregarious and eager to talk about their birds.

Courtesy of My Pet Chicken, LLC

Poultry shows provide space for a sale table and are good places to acquire birds. You can see what they are like as adults and talk with the breeders. Bring cages with you.

Definitions

Hen: An adult female chicken that has laid eggs for six months.

Pullet: A young female chicken, usually less than a year old, or until she has laid her first egg.

Rooster or Cock: An adult male chicken.

Cockerel: A male chicken less than a year old.

Capon: A neutered male chicken. Castrated like steers and geldings, they are more tractable and gain weight better. They are considered a specialty table fowl. Any breed of chicken can be caponized.

A Nankin breeder at a recent show was horrified by some of the birds entered. This rare bantam breed has dwindled, but its charming personality and attractive plumage have caught some breeders' eyes. It is making a comeback.

Showing rare breeds can be a small world. It's not uncommon for all the breeders of a rare breed to know each other. But at this show, someone new had entered some birds.

They were far from the *Standard*, upsetting the breeders who wanted their breed improved. The experienced Nankin breeder was determined to find out who this person was and why they didn't know better.

The birds had been entered by a young girl whose grandfather was tutoring her in raising poultry. They didn't have experience with Nankins, but the birds had charmed the girl and her grandfather was eager for her to pursue the hobby he'd enjoyed.

The Nankin breeder saw a way to encourage them and help his breed. He offered to give them a trio of his birds and work with them on improving the breed, on the condition that they donate their current birds to the local petting zoo.

Nankins are well known for their love of people. The girl and her grandfather were grateful and willing to let the birds go on to a featured role in livestock interpretation.

Breeders are eager to help each other and novices. Never hesitate to ask. As you learn about poultry, you may be in the honored position of offering a hand to others.

Corallina Breuer

These Buff Silkies are healthy and vigorous, good examples of their type. Their offspring would be good stock for exhibition or companion birds.

Corallina Breuer

The color blue attracts attention, but it is a breeders' challenge. It does not breed true, so only some of any given hatch will have the color. These Blue Cochin hens are excellent specimens.

HOUSING

Dennis Harrison-Noonan

Your chicken coop can be attractive as well as practical. Adding flower boxes and other colorful decorative accents makes it part of the landscaping.

Chicken coops can be as elaborate or as simple as your aesthetic sense and your budget dictate. Chickens need protection from predators and adequate shelter from the elements. The coop needs to be accessible so that it can be kept clean. And it needs to conform to local legal requirements. Beyond those basics, it's up to you.

Many chicken fanciers have decided that a chicken coop is the latest fashionable addition to their landscaping. The humble chicken house of the traditional farm has grown up into attractive yard décor. In urban and suburban settings, where neighbors are close by, attractive colors and landscaping make a good impression.

Utilitarian small-production flock owners might not indulge their creative side in housing their chickens, but modern materials offer excellent housing solutions at a reasonable cost.

Many backyard flocks are housed in a chicken house with a small fenced yard where the chickens are protected until they are allowed out of the enclosed yard to roam. They are safe inside the enclosure but aren't locked in the chicken house. The chickens have free access to food and water, as well as to nest boxes to lay their eggs. The chicken house needs to be surrounded by a fenced area where the chickens can be active. The size of this area depends on where else the chickens go during the day. A small enclosure, allowing three to five square feet per chicken, is fine if they get to roam around the yard part of the day.

It came as a great surprise to me that my chickens would put themselves to bed in their own chicken house without me having to round them up. "Going to bed with the chickens" took on real meaning when I left them to their own devices and they calmly followed the sun to bed.

Special considerations need to be taken into account, such as the separation of roosters from fighting breeds. Generally speaking, there are as many ways to house chickens as there are people to think them up.

Hens of various breeds mix well in a small flock. This flock includes an Ameraucana (left) and a Silver Laced Wyandotte (right).

LEGAL REQUIREMENTS

Check with the local governing authority to find out what you need to do to comply with the law. Most municipalities do not require a building permit, but you need to find out what is required in your area. Local ordinances typically cover parameters such as setback and nuisances like offensive odors. Get a copy of the local ordinance and hold on to it. Being in compliance with the letter of the law can help if you run into complaints.

The best way to avoid problems is to be a good neighbor in the first place. If you are new to raising chickens, approach your neighbors first with the idea and reassure them. Keep your chickens clean to avoid odors. Free eggs can also work wonders.

CHOOSING A SITE

The chicken coop and exercise yard need to be located in a well-drained place. Dampness encourages the growth of molds and other undesirable organisms. Chicken droppings are moist and the chicken house needs to stay as dry as possible.

Southern exposure is desirable. It gives the sun the longest time to warm and dry the coop. Traditional farmers often took advantage of the sun to orient their chicken

One day when I had my first chickens in San Jose, California, a visiting neighbor gazed around the yard and said, "Uh, I think one of your chickens is stuck in the dirt."

The furious digging and wiggling she was witnessing was one of the chickens taking a dust bath. From all appearances, this venture is very enjoyable to chickens. They fluff up their feathers and appear to dig their way into the ground. After they have fluffed enough dirt into their feathers, they rub their bodies against the ground. When they have done enough rubbing, they get up and shake all over.

Sand and loose dirt are the best dust-bath materials. Chickens will dust-bathe in whatever is available, but materials such as wood shavings and straw are too coarse to do a really good job. Chickens accustomed to one material may resist changing to a different one.

Poultry scientists have determined that dust-bathing removes stale oil from the feathers, allowing the chicken to replace it by preening. Chickens have an oil gland at the base of the tail. Watch them touch it with the beak and spread oil on their feathers. Dust-bathing removes lice and helps control parasites.

Dust-bathing is a natural behavior. Chicks will start bathing in their starter feed.

Pecking order is enforced at the dust-bath site, but all chickens will have a chance at it. The highest-ranking chicken gets to decide when and where she will take her dust bath, then the others will take their turns.

Corallina Breuer

Dust baths are an important part of a chicken's social life as well as daily feather and skin care. This hen is working particles of bedding into her feathers. Then she will roll around to rub them around, get up, and shake them out. Dust-bathing gives chickens great pleasure.

yards. Windows facing away from prevailing winds, on the south and east sides, provide the best ventilation. Fresh air is important to replace the carbon dioxide of respiration and the ammonia of excreta.

The more vegetation and variety the yard encloses the better. Small trees and shrubs provide much-needed shade since chickens can easily die of heat stroke.

Plan part of the yard for dust bathing. Chickens love a good dust bath and it's important in maintaining the quality of their feathers. Allow enough space for several chickens to dust bathe together. Sand or loose soil is best, so the chickens can work small particles down to their skin. If the site doesn't offer a natural dust-bath site, make one by building a low-sided box and filling it with sand. I once used a child's wading pool, which worked well.

BUILD OR BUY?

If you have a rural property, you may have an outbuilding that can be converted into a chicken house. Any shed, barn, lean-to, or other preexisting structure may be adapted to becoming a chicken coop. Children's playhouses have gone on to become chicken houses after the kids are grown.

Reusing materials is one of the satisfactions of the thrifty. Make sure any materials you reuse to build a chicken coop are safe, without any protruding nails, sharp edges, or loose wire. Minor injuries to chickens or yourself can get infected and cause major problems Consider, too, whether you want to use treated lumber or if the material

Courtesy of Omlet USA

Chickens enjoy the best of indoors and outdoors with simple and elegant Eglu. Children can manage it with some help from parents. Additional protection is needed in cold climates.

The Eglu is a good solution for suburban chickens. It's safe, easy to move and clean, and comes complete with chickens, if you so desire.

Courtesy of Omlet USA

has been exposed to hazardous chemicals that could be harmful to the birds or their meat and eggs.

Insulation and a good vapor barrier help the coop stay dry. Doors should open inward. Avoid windows or other structures more than four feet high that could be used as perches or roosts. Giving the floor a slight slope can make it easier to hose out.

An English company is now making the Eglu, a simple plastic shelter and secure run for suburban chickens, available in the United States through Omlet USA. Through a cooperative arrangement with rare-chicken hatchery Murray McMurray, the company will even send a starter set complete with chickens. At $740 (chickens, $10 each, are extra), it may be the perfect no-fuss solution.

CHICKEN TRACTORS

Another option for housing poultry is the chicken tractor, which got its name because it works in two ways: as a safe, healthy pen for chickens and as a way to turn over the soil and prepare it for planting. These structures are also called arks, mainly in England. Chickens are amazingly assiduous in scratching up weed seeds and bugs, turning them into fertilizer, and working it into the ground. A chicken tractor takes advantage of their natural behavior.

A chicken tractor is a secure, enclosed pen without any bottom, so the chickens have free access to the area underneath to peck and forage. It may have a small henhouse attached, where the birds can roost and sleep at night, or they may spend the day in the tractor and go back to another place to sleep. The tractor may have wheels on it so it can be moved from place to place. This can be done with or without chickens inside.

A chicken tractor should allow at least one square foot inside the house per hen and three square feet in the run. A chicken tractor for twenty-five hens could be 5 feet wide and 20 feet long, with a 5x5-foot house.

Chicken tractors have a lot of advantages for both parties—the chickens and their keeper. Even if you have a chicken house and an enclosed yard, you may want a chicken tractor for some parts of your property.

Chicken tractors need to be completely covered, as the chickens will not have a place to take cover from hawks and owls. They need to be sturdy and the sides must come down

to the ground. You need to keep an eye on them, to stop any predators from digging underneath to get to the chickens.

Chicken-raisers Andy Lee and Patricia Foreman have written several guidebooks on pasturing poultry, *Chicken Tractor* (1994) and *Day Range Poultry* (2002). Their day range system includes sturdy movable housing to protect the birds at night, set in pasture enclosed by electric netting to keep the chickens in and predators out. Their systems are directed at the hobbyist/entrepreneur who raises poultry for profit, as well as for fun. The welfare and well-being of the chickens are important considerations.

FENCING THE RUN

The chicken run must be completely fenced around the perimeter. Fencing the top isn't essential when chickens can

Corallina Breuer

This chicken house is on wheels, so it can be moved to other locations to take advantage of local conditions such as extra sunlight or better drainage.

Corallina Breuer

These Buff Laced Polish hens have a safe fenced run to enjoy fresh air and sunshine. The chicken house offers protection from inclement weather and predators.

take refuge in an adjacent coop, but it does offer additional protection against hawks and owls. Burying wire fencing a foot deep, bent outward, will foil predators, such as coyotes, raccoons, and dogs trying to dig under the fence and get to your chickens.

Electric fence can be run around the outside of the pen, about four inches off the ground to discourage predators. This was the only solution I found to a nasty dog problem I once had.

Snakes and rodents will also look for ways to get in. Pouring a concrete floor or raising the chicken house on blocks or posts can keep your chickens safe.

SIZE

The size of the chicken house and exercise yard depends on how many birds it needs to house. Figure a minimum of 1.5 to 2 square feet per chicken inside the chicken house and 8 to 10 square feet of yard outside. Bigger is better. Overcrowding chickens results in nasty problems of pecking and even cannibalism. They will not thrive. They need space to flap their wings and settle the endless social jockeying for which chickens are known.

Chickens are perching birds and do better if they have a perch to sleep on. Inside the house, figure six to ten inches of perching space per chicken. Roosting can become competitive, with the top perch winning the most desirable roosting spot. Avoid problems by having all roosts at the same level. Again, more is better than less, as chickens low on the pecking order are sometimes chased off perches and they need alternate places to go. Allowing them to sleep on the floor or in nest boxes exposes them to unsanitary conditions, as the litter is inevitably soiled.

One nest box for every four to five hens is adequate. Since chickens, like other birds, are always trying to lay

Corallina Breuer

Chickens prefer the highest possible roost. Some can fly that high, but others need a ladder to get into boxes off the ground. Lighter breeds are generally better flyers.

This Ameraucana hen happily lays her eggs in plastic nest boxes. Chickens may choose the same nest box every day or be willing to use other locations. The pecking order governs nest choice and timing.

Corallina Breuer

enough eggs to get a clutch—that magic number that looks like the right number to hatch—they usually want to lay in a nest their sisters are using. If you have a preference for where they lay their eggs, put an artificial nest egg there and they will accommodate you.

It's not uncommon for two or more chickens to get into a nest together. Unless they are breaking eggs or pecking each other, leave them be. They may as well nest where they are comfortable.

Corallina Breuer

These bantam Mille Fleur d'Uccle hens have no trouble flying up to lay their eggs in the top next boxes.

LIGHT

If you plan on influencing your hens to lay year-round, as most farmers do, electric light is necessary. Sixteen hours of light a day will keep them laying throughout the winter months.

Electric light will also add a small amount of heat for cold climates. Chickens are well-feathered and typically manage cold weather well. They generate a lot of heat on their own. However, a small amount of artificial heat when temperatures are below zero can help prevent large combs from freezing. A frozen comb is stressful for the bird and it will never grow back, ruining the bird's show potential. In cold climates, fanciers often choose breeds with rose or pea combs to avoid the problem.

Courtesy of My Pet Chicken, LLC

Chickens typically manage cold weather well, insulated as they are in feathers. However, combs are susceptible to freezing, and breeds with rose or pea combs do better in bitterly cold climates.

HEALTH AND CLEANLINESS

Chickens need fresh air and sunshine to thrive. Their house must have good ventilation. They need to get out in the sun but have places where they can find shade. They need to be able to stay dry when it rains. This is especially important for Silkies, who, with their hairlike feathers, can get soaked and chilled. Chickens with conventional feathers repel water, a challenge when it comes to bathing them! Hard-feathered breeds like games repel water best.

The structure used as your chicken house must be secure against whatever climate conditions prevail in your area. Mild climates require less protection against cold, but no location is without its dangers. Recent hurricanes in the South literally blew the feathers off chickens.

Litter on the chicken house floor will absorb moisture from the droppings. Chopped straw, sawdust, dry leaves, and cedar shavings all make good litter. You should haul out soiled litter and replace with clean at least once a week, to protect the chickens from ammonia in the air they are breathing and to avoid excessive bacteria. Cleaning every day isn't too often. Keeping the chicken house clean also reduces odors, a consideration if you live near your neighbors.

Design easy access into the chicken house. A full-size gate or a roof that opens for easy cleaning helps. Easier access means cleaning gets done more often. Consider whether you want your children to be able to manage this chore.

Cleaning the chicken house may not be your favorite chore, but it's one of the best ways to spend low-key time getting to know your chickens and letting them get to know you. It's really not unpleasant.

If your flock should suffer an infestation of parasites or contract a disease that requires decontamination, the process will be made easier if convenient access is built in. Paint roosts with used motor oil to control mites, which live in crevices.

Corallina Breuer

Chickens need easy access to their nest boxes, otherwise they will lay eggs in places of their own choosing and you will have to search for them.

Corallina Breuer

A wire door can provide additional security while providing fresh air and good circulation. Match housing to your local climate and individual needs.

PREDATORS

Predators, both aerial and terrestrial, are the worst problem you will confront while raising chickens. Raccoons, possums, mink, snakes, and even neighborhood dogs will prey on chickens. From the sky, owls and other raptors prey on chickens, night and day. Waking up to find the entire flock devastated has disheartened more than one chicken farmer. Nearly every chicken owner loses birds to predators occasionally, but being vigilant reduces your losses. A completely predator-proof facility is the poultry owner's answer to perpetual motion. It hasn't been achieved yet.

A secure construction is the first line of defense. Keep fencing in good repair and leave it locked when you aren't around. Raccoons are particularly clever and persistent about breaking and entering. I watched a hawk pluck at the chicken wire covering my coop in a very determined way; with time, she probably could have gotten in.

Sometimes it is difficult to tell what is raiding your henhouse at night. Motion-sensor lights may help. Burglar alarms can scare off the intruder or at least alert you that it has arrived. Installing a web cam is within reach of many owners. Watching the web cam video can show you what is sneaking in and how it is getting there.

A watchdog may be your most reliable ally in the struggle against predators. Any good dog will do, so long as she understands that protecting the chickens is her job. Serious breeders rate the Anatolian Shepherd as the best defender of their stock. A breeder in California reported that her Anatolian Shepherd even held off a mountain lion.

Dealing with Predators

Shooting raccoons, opossum, and rats is the traditional method of controlling predators. The challenge is getting them in your sights. Staying up all night trying to see in the dark is hard on the keeper.

By poisoning pests you risk poisoning the chickens, pets, and children. A three-chambered trap, with the poison in the third chamber, avoids leaving the poison in the open, but other animals (including your chickens) might eat the poisoned carcass.

Leg-hold traps can be effective. Cat food is a tempting bait for many predators.

Live traps can trap your marauder, which can then be killed or relocated. Be thoughtful about relocating a problem animal. Check with local game wardens about finding a location where it won't simply become someone else's problem. Make sure, too, that it's far enough that it won't make its way back to your chicken coop in a day or two.

Older cookbooks have recipes for opossum and raccoon. Trapped opossum can be fed on milk for ten days before killing and eating. Make sure to remove the small red glands in the back and under each foreleg. The meat requires extra processing to reduce its gamy flavor. Blanching it in several changes of water before roasting helps.

Craig Russell of the SPPA prefers young raccoon to older, tougher animals. Young raccoons make good pot roast with mixed vegetables and horseradish on the side. "Even good eating won't replace rare stock, but it may give you some satisfaction," Russell says.

Older animals can be slow cooked, deboned, and chopped or ground to be served as hot sandwiches with gravy or barbecue sauce.

Old-time poultry keepers grind up the whole animal and feed it to their flocks, a sort of final turn in barnyard justice.

Courtesy of the National Conservation Training Center, Publications and Training Materials

Chickens are vulnerable prey that require protection from wildlife. Predators such as this raccoon are formidable contenders. They are untiring in their determination to have your chickens for themselves.

CHAPTER 6

• •

FEEDING

Chickens are omnivores—they eat all kinds of food, both plant and meat. They are resilient and able to thrive on a wide range of diets.

Like all critters, they do better on a diet that meets their nutritional needs than one that falls short. They lay more eggs and are more resistant to illness if they are getting a good ration.

As domestic animals, chickens rely on their caretakers for sustenance. If they are on pasture or free-range, they will seek out the seeds and bugs the land offers. Any given piece of pasture may not have everything chickens need or may have it only at some times of the year, such as when caterpillars are hatching.

Confined birds are entirely dependent on their keepers. Providing a balanced, nutritious diet is the first step to enjoying your chickens and a successful egg and meat operation. Confined chickens are usually fed twice a day.

Good physical condition, strong general health, and regular egg-laying are reliable indications of good nutrition. Healthy chickens drop healthy poop. Chickens excrete solids and liquids together, so droppings should be moist with some brown solids. Unusual colors, such as brownish orange or yellow, can indicate a health problem. Runny droppings in response to dietary changes suggest something is not agreeing with your chickens. Excessive grain can cause

runny droppings. Runny droppings with blood can indicate worms. Pay attention to the general health of your flock.

Chickens vary in size. Although commercial Cornish/Rock crosses are slaughtered at six to eight weeks of age, growing to full size takes much longer. Some breeds grow more slowly than others but should be vigorous and steadily gaining weight. Heirloom-breed males may take as long as a year to reach their full bone development and weight.

Chickens can also get too fat, which interferes with laying. Hold the bird in your hands and feel for a fat pad between the legs. If it's cushy, they are too fat. Gradually reduce the amount of grain or other carbohydrates in the diet.

Although chickens are adaptive and enthusiastic about all kinds of food, their owners have preferences and sensitivities that affect their choices of feed. You may want to avoid animal products in commercial feed or make your own feed. Dr. Charles Everett of South Carolina grows his own grain for scratch feed. He saves seed from the amaranth and sunflowers he grows to perpetuate the heirloom breeds he and his chickens prefer. He has developed his own strain of gourdseed corn and also purchases whole oats and crimps them in his own feed mill. These feed options are available to all small-flock keepers, if you have the inclination.

Corallina Breuer

This mixed flock is fed grain and ration in the metal feeder on the ground. Sufficient space assures that even the birds at the bottom of the pecking order will have an opportunity to eat.

Commercial feeds are prepared as mash, crumble, pellet, and scratch grain. Crumble and pellet preparations provide more reliable nutrition, as the processing combines all ingredients equally. Heavier ingredients in mash may migrate to the bottom, resulting in uneven nutrients.

Courtesy of Nutrena Feeds

FEED

Chickens need a diet that supplies carbohydrates, oil, protein, vitamins, and minerals. They can get all these nutrients from grain, greens, and animal or insect sources.

Scratch feed is a mixture of whole and cracked grains such as corn, wheat, and sorghum. Feed stores carry it in large sacks, with the ingredients listed on the label. It is the traditional supplement to barnyard chickens that range to forage for greens and insects and receive plenty of kitchen scraps; it does not give chickens a complete diet.

Food closest to the source is always preferable. Whole grains are better than cracked grains. Manufactured feed is inevitably older than feed you make yourself. Chickens benefit from all kinds of grains: alfalfa seed, sunflower seed, wheat and wheat germ, sesame seeds, oats, rice, rye, barley, millet, flax seed, amaranth, and others.

Whole grains and seeds contain oils necessary for good feather condition and carry important vitamins for good health. Commercial feeds have oils added to replace those lost in processing.

Commercial feed is a good way to start with chickens. As you proceed, you can learn more of the refinements of poultry nutrition and, if you like, move on to making your own feed. Many companies and some local feed mills make their own formulations.

Chick starter has around 20 percent protein. Chicks are usually kept on starter for at least sixteen weeks, or until about half of them are laying eggs. Their owners can then transition them to a grower ration and then to layer crumble or mash.

Rapid growth may be a primary goal for commercial operations, but growing your small-flock chickens too fast can result in problems. Commercial chickens have been selected for rapid conversion of feed into meat. Traditional heirloom breeds grow and develop more slowly.

Chickens, especially cocks, that grow too fast may develop a leg deformity known as "going down on the hocks." It is a permanent defect that makes them unusable. The leg problem may be caused by inadequate absorption of B vitamins. Brewer's yeast or nutritional yeast is rich in B vitamins and may help the digestion to absorb these important nutrients better. Experienced breeders recommend adding brewer's yeast to the chicks' water during their thirteenth week of life. If any leg weakness appears in the following week, supplement with brewer's yeast for another week.

Adult crumble, another complete manufactured feed, ranges from 16 to 18 percent protein. It also contains calcium, which mature hens need to make eggshells but which can permanently damage chicks' kidneys.

Types of Feed

Scratch feed: A mix of whole grains including corn, oats, amaranth, wheat, rye, millet, sorghum, and others. Provides carbohydrates, B vitamins, phosphorus, and some protein. Often scattered on the ground so chickens will scratch for it.

Starter: Mash feed formulated for chicks, ground fine enough for them to eat it but not so fine that it is powder. From six to fourteen weeks, it should be around 17 percent protein. From fifteen weeks to twenty weeks, 14 percent is sufficient.

Pellets: Compressed ground-feed ingredients, blended so that nutrition is uniform.

Grower: Mash feed formulated as transition from chick starter to adult feed. For production birds, broiler grower mash is 19 percent protein. Pullet grower mash is 16 percent protein. Protein may come from soybeans, fish meal, or other sources. Not more than 1.2 percent calcium.

Layer crumble: Pelleted feed blend broken into smaller pieces for hens that are laying, including 2.5 to 3 percent calcium and 15 to 19 percent protein.

Mash: A blend of feed ingredients ground fine but not to powder.

Brewer's yeast: A good source of B vitamins (except for B12), other vitamins, sixteen amino acids, and many minerals.

Chickens require a small amount of salt in their diet. Commercial feeds include the proper amount of salt. If you are making your own feed, include a small amount of salt—less than 0.15 percent. More will cause soft-shelled eggs. Be mindful that a lot of leftovers contain salt. Chickens can die of excess salt.

Alternatively, if you provide your chickens with unsalted feed and give them access to a separate source of salt, they will manage their own salt intake. One option is dried kelp. If you live near the ocean, you can collect kelp from the beach, dry it, and offer it to your birds. It will supply all the minerals and salt your birds need.

All feed must be kept clean, dry, and safe from vermin. Chicken feed is otherwise known as rat feed. It spoils easily if it gets wet, and moldy food can kill chickens. Plastic garbage cans with secure lids make good feed containers.

GRIT

Chicken digestion is not like mammal digestion. They have a crop filled with gritty sand that grinds up their food along the way to digestion. They need a regular supply of grit in their diet to allow them to digest their food. If they are on pasture, they will pick up grit from the ground. If they are not able to have free access, provide them with a dish of grit.

PROTEIN

The source of protein—whether animal or vegetable—in commercial feed is listed on the label. Protein from animal products is recovered from the processing of meats such as pork and other poultry. Protein from vegetable sources also fulfills the basic nutritional needs of chickens.

Chickens enjoy insects of all kinds. Small-flock owners have satisfied their chickens with everything from trapped yellow jackets, attracted to restaurants by nontoxic aromatics, to fly maggots and earthworms. One of my chickens was a regular companion when I weeded the garden, taking advantage of the tasty worms I dug up. Snails, a problem in some areas of California, were so plentiful that my chickens became blasé about a new shipment.

My gardening neighbors used to bring my chickens pails of snails to get rid of them.

Corn, wheat, and oily sunflower seeds supply protein. Roasted soybeans are a good source, though not all chickens will eat them. Fish and fish meal, worms, and insects of all kinds are tasty treats. Each has its own balance of nutrients, so proceed with caution.

Chickens on pasture will feed themselves well, so long as the pasture provides enough bugs, worms, and plants. Rotate chickens on pasture and give the land time to recover before using that pasture again.

GREENS

Greens provide chickens vitamins and minerals as well as fiber. They enjoy grass and weeds of all kinds and will happily eat up your garden if they can get into it. Any garden waste makes an interesting challenge for chickens who are attracted by new things in their yard.

Green kitchen cuttings are wonderful chicken food. Get additional greens from the local grocery store. Ask for produce trimmings. Some produce managers will give them to you free or charge a small amount. You can usually get large bags of trim that your chickens will greet with delight.

Clean grass clippings and weeds make good greens. Some people sprout grains such as wheat, oats, barley, and rye for their chickens. They can be soaked and sprouted on trays lined with moist paper. Lori and Gil at Sproutpeople found their chickens enjoyed all kinds of sprouts. They liked wheat grass in the winter. Their Middle-Sized Bird Blend is suited to chickens.

Robert B. Edwards, USDA ARS Image Gallery

Many kinds of corn make good chicken feed. Chickens like all kinds of corn and will happily eat sweet corn by the ear for a treat.

Corallina Breuer

This Barred Rock bantam cock is happy to eat bugs from the garden. He'll also dig up the plants, so put chickens in the garden only after the harvest is complete.

Bruce Fritz, USDA ARS Image Gallery

Black sunflower seeds are preferred for nutrition, though chickens like all kinds of sunflower seeds. A sunflower head hung in the run can encourage exercise and give curious chickens an interesting activity.

Grow Your Own

Dr. Charles Everett, a Baptist minister in Lugoff, South Carolina, balances the layer crumbles he feeds his chickens with homegrown scratch feed made of corn, amaranth, sunflower seed, and oats.

He avoids modern hybrid corn because of its lower protein level, less than 9 percent, in favor of colored heirloom varieties such as Bloody Butcher, Ohio Blue Clarage, Texas Gourdseed, and Warner. Although he finds their yields lower, the protein content is around 9.75 percent. By crossing gourdseed corn, the kind grown in the South prior to the introduction of hybrid corn in the 1850s, with colored varieties, Everett developed a variety that is easily shelled, drought-resistant, and has higher protein.

Everett sows the amaranth thickly, thinning the six-inch plants and feeding those to his chickens. The heavy seed heads are easily harvested and the seeds require no processing.

Everett grows two kinds of sunflowers for seed: Russian Mammoth and Hopi Purple Dye. He credits the higher oil content of the Hopi Purple Dye seeds for his birds' shiny feathers. Black oil sunflower seeds, available in fifty-pound sacks, are a good substitute. Like whole oats, they need some processing to be digestible to chickens.

Everett doesn't grow his own oats, but whole oats are easily available as feed or may be purchased from a local farmer. They need to be crimped to be digestible for chickens.

Basic ratios of ingredients are 10:1 corn to amaranth; 5:1 corn to sunflower seeds; and 2:1 corn to oats. Everett adds five pounds of oyster shell and half a pound of salt to every hundred pounds of feed.

Everett feeds seven parts scratch to one part laying mash, giving the birds as much as they can clean up in ten minutes, morning and night.

"There is a satisfaction that comes from being able to grow and grind your own supplemental scratch grains," he says.

Corallina Breuer

This group of Sumatras has everything they need: a grassy yard to browse for greens, grit, and protein in the form of bugs and worms. The sandy spot makes a good dust-bath spot.

Chickens forage for their food on the ground. It's a useful evolutionary strategy for life on the forest floor, but commercial crumble, mash, and pelleted feeds are not designed for chickens' most basic natural instincts. Chickens shoveling with bills and scratching with feet in commercial feedstuff make a mess.

Kermit Blackwood set out to solve the problems he observed. An animal management intern for the Wildlife Conservation Society and the general curator of Marlboro College's Life Science collection in Vermont, Blackwood works with green jungle fowl, Bornean white-tailed pheasants, and Congo peafowl. Nutrition proved to be one of the most challenging issues of working with these rarest of wild Galliform bird species, such as wild high-altitude adapted Himalayan pheasants and domestic heirloom strains of Japanese Jittoko and Minohiki fowls, French Marandaise, and Oceania's Rapanui and Mapuche fowls.

Blackwood observed that vegetable-based and grain-based feedstuffs move so quickly through a chicken's digestive system that the food isn't fully digested. Consequently, the birds produce copious amounts of acrid, partially digested droppings. Although the fowl are eating a lot, they aren't utilizing all the nutrients. Their foraging behaviors end up contributing to the inefficiency of the feedstuff. Ultimately, the birds may consume as little as 35 percent of the feed put out for them.

Commercial feed formulations are generic in nature. Chickens are adaptable and thrive on a wide variety of diets, but some breeds do significantly better with different formulations, especially heritage breeds with long and productive life spans.

The foraging behavior of chickens mixes disintegrated feedstuffs with dirt, dried fecal material, feather dander, and other undesirable matter. The birds breathe the dust. Despite careful management, a cycle of infection and disease sets in. Low-cost chicken feed may mask other costs—in reduced production and vitality due to chronic poor nutrition and illness.

Blackwood and directing manager Charles Clour set up Resolve Sustainable Solutions and developed Forage Cakes to address these issues.

The company produces three product lines formulated to American Zoo and Aquarium Association guidelines for sustainable agriculturalists, alternative livestock managers, and private aviculturalists.

The company's Forage Cakes are essentially giant granola bars, formulated as nutritional/behavioral supplements for chickens and other fowl. Used as directed, Forage Cakes will help captive birds better utilize their entire diet.

They are made primarily from byproducts produced by sustainable agriculturists. Cranberry seeds and pulp byproducts of juice processing, vegetable pomace from processing V-8, and organic bakery leftovers are mixed together with beef gelatin, which glues the pressed cake together like wild bird seedcakes. Embedded treats such as nuts, seeds, grains, and flower petals encourage natural foraging behaviors, which is highly stimulating for confined chickens. Crustacean meal from krill and crab adds an ideal source of easily digestible protein for all poultry species, including domestic breeds and wild fowl in every growth phase. Krill is the primary source of animal protein in all formulas of Forage Cakes. Freshly ground spices are naturally antibacterial, fighting off infections and discouraging parasites, while making the Forage Cake unpalatable to vermin. Cranberry seeds have antioxidant properties. Ethiopian teff seed, grown on farms in Idaho, is high in trace minerals and protein. These natural ingredients boost the birds' immune systems against infection and disease.

The company is currently developing a line of specialty formulas specifically to meet the distinct needs of heirloom breeds.

"Our generation is the last one to have the opportunity to revolutionize the way people feed and maintain their poultry," Blackwood said. "Forage Cakes will do that."

Courtesy of My Pet Chicken, LLC

This White Leghorn runs to catch a juicy bug. Chickens must be alert to their surroundings and able to act quickly to succeed as foragers.

CALCIUM

Greens contain plenty of calcium, another important nutrient for adult chickens. Chick starter does not contain calcium, as it can interfere with their development. Never put adult crumble, oyster shell, or other calcium feed out for chicks. When pullets get ready to lay eggs, though, they need it. Eggshells are made of calcium and hens can get depleted if they aren't getting enough in their diet. Commercial layer crumble provides adequate calcium and additional sources are unnecessary. For hens on a diversified diet, however, you should supply their calcium needs with crushed oyster shells in a separate feeder.

FEEDERS AND WATERERS

Wall-mounted or hanging feeders in more than one location help entertain confined birds. Having more than one feeder also helps birds low in the pecking order get enough to eat.

Mash and crumble can be messy, spilling out of the feeder and creating an unsanitary condition in the pen. Feeders with an extra lip over the feeding trough or protected by a grill reduce spilling.

Waterers should be kept clean. Chickens prefer clean water and will avoid fouled water, even to the point of death. Waterers should be washed frequently and rinsed with bleach every week or two.

For chickens, eating takes up a significant amount of the day. They like to enjoy their food. Giving them ways to forage or work for their food adds enjoyment to their lives. Hanging greens around their pen encourages them to get some exercise jumping for their food. They will also eat from your hands. Watch them scratch for food and thrill to the discovery of a tasty bug. They are endlessly interesting.

This indoor flock of Leghorns at the University of Wisconsin manages with two hanging feeders. Hanging feeders are designed to reduce waste, but some feed inevitably lands on the ground.

Corallina Breuer

Medicated Feed

Medicated chick feed contains such a low dose that it allows them to develop and recover from infections, acquiring their own natural immunity without being killed by the infection. They aren't laying at this point, so antibiotics aren't getting into the eggs.

Coccidiosis is devastating to chicks. It is caused by a protozoan parasite, the eggs of which are commonly found in soil. If you have only a few chickens, keeping them scrupulously clean will help protect them.

Using antibiotics appropriately to fight infections is different from constantly medicating. Antibiotics are one of the great discoveries of our age. Withholding them from sick chickens is unconscionable. Being pure on this issue makes less sense than being wise.

Homemade Feed Recipe

This regimen has been successfully used by traditional poultry keepers for many years:

Start chicks on a mixture of starter crumbs and finely chopped hard-boiled eggs for two weeks. This is a good use for infertile hatching eggs.

At two weeks of age, add chick grit, wheat, and finely cracked corn. Gradually decrease the amount of chick starter until it is only about half the ration at three months of age. Then mix of a ration of:

- 500 lbs. wheat
- 100 lbs. corn
- 100 lbs. barley
- 100 lbs. oats
- 100 lbs. black oil sunflower seeds
- 100 lbs. buckwheat

Mix half and half with grower crumble starting at three months of age. At six months, transition to half and half layer pellets.

For the birds you plan to breed, replace a quarter of the layer pellets with Calf Manna, Purina Animax, or other supplement six weeks before you start collecting hatching eggs. Many commercial feeds contain an antibiotic, usually Amprolium, to fight coccidiosis. Some also contain Bacitracin. Avoiding subtherapeutic doses of antibiotics may be one of the reasons you want to raise our own meat and eggs, but consider the issue before deciding against it for your flock.

Corallina Breuer

Gravity-controlled waterers have a valve system that assures an even supply of water to all levels of cages. Otherwise, the water pressure would overflow the bottom cages without supplying the ones on top.

Corallina Breuer

This hanging waterer automatically replenishes the drinking trough around the edge. As the water level drops, the waterer gets lighter, releasing pressure on the supply line. Water flows down over the plastic dome and into the trough.

FLOCK
MANAGEMENT

Corallina Breuer

This mixed flock includes Silver Laced Wyandottes, Golden Ameraucanas, and other breeds. Without a rooster, the eggs will not be fertile but can be sold as free-range eggs.

This chapter pulls together general principles and practices for sound management. Beyond the basics of good nutrition, fresh air and sunshine, and protection from predators, there are many ways to manage a flock. Chickens are adaptable and hardy.

How you manage your flock depends on your physical circumstances as well as your personal inclinations. Chickens close to neighbors in suburbia live under different constraints from chickens in rural areas with space to spread out.

Within those basics, chicken owners over the years have devised systems that work for them. Adapting their ideas to your circumstances will work for you. If chickens become a bigger part of your life, you may find yourself changing those circumstances to include more land for your growing avocation.

ORGANIC POULTRY

If you are raising chickens for either meat or eggs to sell to the public and intend to designate them as "organic," you will have to meet the standards of the National Organic Program. Organic meat and eggs come from chickens that are given no antibiotics and are fed organically grown grain and other food. Federal regulations prohibit the use of hormones in all poultry production, so not having hormones

is not unique to Certified Organic poultry. Organic food is produced without using most conventional pesticides, fertilizers made with synthetic ingredients or sewage sludge, bioengineering, or ionizing radiation. The USDA has a Certified Organic program and many states have their own programs.

Other terms, such as "natural" and "free-range," are less precisely defined. Generally, "natural" means that artificial ingredients or coloring have not been added or injected into the bird at the processing plant. "Free-range" or "free-roaming" means that the birds have access to the outdoors. "Cage-free" means the birds are not confined in cages but grouped together in an enclosure.

Unfortunately, some producers have used marketing and advertising to stretch the usually understood meanings of these words to apply to conditions that would disappoint the consumer. "Free-range" chickens may have access to the outdoors, but it may be bare ground that doesn't offer shade or interesting things to scratch for. The chickens may not bother going outside.

Chickens enjoy foraging for part of their diet by scratching for seeds and bugs. If you intend to market birds raised this way, make sure your customers understand the care the chickens are getting.

Consult with local authorities, such as farmers' market directors, to determine what is required to market your products through them. County extension agents will also have information on marketing to the public.

PASTURED POULTRY

Raising poultry on pasture allows chickens to be part of a productive cycle of farming. As they forage, they turn over the soil with their digging and consume weed seeds. The cycle is completed when their manure works its way into the soil and improves it.

Pastured chickens need to be protected from predators. They need to be moved to fresh pasture as they deplete the land they are working. Pasture is clean and healthy for chickens. They are able to express normal chicken behavior when they are on pasture.

Ten to fifteen chickens can work over an acre of land in a day. In planning an enclosed chicken yard, allow 10 to 15 square yards per bird. The chicken yard may be divided, keeping chickens on one side at a time, giving the other side a chance to recover.

Chicken tractors allow chickens to perform the valuable service of cleaning up weeds and other unwanted vegetation while digging up the ground and fertilizing it. These portable pens give small flocks safety and comfort with the byproduct of compostable manure. For more information about chicken tractors, see Chapter 5: Housing.

A useful resource for flock keepers who want to pasture their chickens is *GRIT!*, a bimonthly newsletter published by the American Pastured Poultry Producers Association.

Andrew Zimmerman

Chicken tractors can take many forms. The unifying principle is that it be movable, so chickens can work over different sections of land. These hens are protected from predators and can take shelter under the tarp.

Composting Manure

Soiled litter is an excellent start to rich compost. It needs some handling to bring it to its best.

Chicken manure is high in nitrogen. The ideal compost recipe is about thirty parts carbon to one part nitrogen, so the manure needs to be mixed with additional brown compost. Green material includes grass clippings, weeds, and kitchen trimmings. The general rule is one part brown to two parts green, but chicken manure is so rich in nitrogen that a one-to-one ratio or even two-to-one will make good compost.

The compost should be in a bin at least one cubic yard where you can mix the green material and soiled litter together and wet it down. It will cook itself. The internal temperature should reach 130 to 150 degrees Fahrenheit. Let it cook there for three days. Then turn the pile, moving the material in the middle to the edges and the material at the edges to the center. Cook one cubic yard of compost at least three times this way.

Hotter is not better. Over 160 degrees will kill the beneficial organisms you want.

After you are sure it has all cooked adequately, a week or so, cover it loosely and let it sit for six or eight weeks. It should be dark, crumbly, and sweet-smelling when it is ready to go on the garden.

Two bins allow you to have one collecting soiled litter and one with the previous batch curing. You may want to pile the cured compost somewhere convenient to the garden.

University of Wisconsin–Madison Center for Integrated Agricultural Systems

This chicken tractor is part of a pastured poultry production operation. Keeping the tractor short conserves materials and makes it easy to collect eggs. Light weight is important to make it possible for one or two people to move the tractor.

University of Wisconsin–Madison Center for Integrated Agricultural Systems

Chicken tractors are well suited to small- and medium-sized diversified farms. Feed costs are reduced because the chickens forage for part of their diet.

BARNYARD POULTRY

Allowing chickens to roam free is romantic but seldom practical, for the people or the chickens. Chickens at liberty are subject to predation by any number of wild and domestic critters.

Chickens enjoy exploring and will get into everything, including your garden. If you decide to let your chickens roam free, you will have to make a decision about fencing the garden.

Chickens are social birds. A single chicken kept as a pet will be lonely. Even if you are tentative about getting started with chickens, plan on purchasing at least half a dozen. It gives you and them some margin for failure and mistakes.

URBAN POULTRY

Because of chickens' growing popularity, many cities have addressed the issue of poultry within city limits. Raising poultry in town is usually legal but regulated for urban dwellers. Typically, roosters are not welcome. Housing for poultry is carefully regulated to conform to urban expectations. So long as you observe the local ordinances, your chickens are welcome.

Portland, Oregon, and Madison, Wisconsin, both allow poultry within their city limits. Before Madison formally regulated chickens, local enthusiasts who raised chickens considered themselves the Chicken Underground.

Modern developments may add restrictions on livestock, including chickens, to property deeds. Read carefully to understand what is allowed on your property. Such restrictions are legal but may be enforced differently. You may be able to reach agreement with your neighbors to allow you to raise chickens in compliance with less restrictive local laws. Be aware that a complaint could trigger a visit from the property owners' association representative and a request that you get rid of them.

Harvey Ussery

This mixed flock is managed on deep litter made mostly of oak leaves on a Virginia farm. The hens are (left to right) Old English Game, Silver Gray Dorking, Kraienkoppe, and Welsummer. The rooster who shares their coop is an experimental crossbreed.

DEEP LITTER

In climates that have severe winter or summer weather, chickens can be kept on deep litter during part of the year and outdoors the rest of the time. Deep litter removed annually produces fertilizer that requires no further composting.

Deep litter requires a dirt floor. The litter should be 6 to 12 inches deep to start. The pen should allow 4 to 5 square feet of floor space per bird. Over time, chickens will break down the litter and reduce it to dust. Additional material can be added to keep the litter deep.

Deep-litter systems are based on the natural decomposition of organic material, combined with chicken droppings and the chickens' natural instinct to scratch the litter. The mixture composts and breaks down chicken manure.

Deep litter maintains a relatively even temperature, generating heat that takes the chill off during cold months and giving chickens a cool place during hot months. It's a natural habitat for bugs that chickens scratch up and eat. Estimates suggest that up to 100 percent of a flock's protein needs can be met by the critters they eat from deep litter.

The litter itself can be any kind of dry organic material, such as fallen leaves, wood chips, sawdust, dry grasses, groundnut shells, chopped corn cobs, tree bark, and chopped straw. The carbon combines chemically with the nitrogen in chicken manure in the composting process.

Deep litter makes good dust-bath material. Chickens will happily work it into their feathers. Chickens kept on deep litter have few problems with lice and mites and the incidence of coccidiosis is also reduced.

The litter should be changed annually. The composted litter is ready to be used as garden fertilizer, since it has around 3 percent nitrogen, 2 percent phosphorus, and 2 percent potash. Poultry litter has about three times the levels of compounds as cattle manure.

Corallina Breuer

These Cornish/Rock crosses are the hybrid that becomes chicken on the American table. They are raised indoors on an all-in/all-out management system.

Corallina Breuer

These Cuckoo Marans are living in a single-breed group. They have access to the outdoors as well as an indoor shelter. All fertile eggs from this group will be suitable for hatching.

Courtesy of the USDA

A compost pile can be managed in a small area. It provides a convenient way to recycle kitchen waste, garden waste, and poultry litter. The natural decomposition process turns these materials into rich fertilizer for topsoil.

Although deep litter requires less work than conventional hen houses, some cautions need to be observed. Wet litter can ferment and become contaminated. Avoid excess water from leaks in the roof, leaky pipes, dripping faucets, or sloppy waterers. The litter must be kept dry, but not parched, for the composting to work. The normal moistness of soil is about right. Peat moss can absorb water from litter that is too wet.

Litter has to be turned frequently. If the chickens aren't turning it well enough or the litter is compacting in places, turn it with a fork, then let the chickens get back to their work.

Good ventilation is essential. The composting litter requires oxygen to make the reaction work. The chickens work the oxygen into the mixture.

Deep litter should not have any objectionable odor. If you smell ammonia, the nitrogen is not being matched by adequate carbon to decompose. Add more litter and mix well.

Larry Rana, USDA

Large poultry production facilities can also use composting effectively, reducing the impact of waste disposal. This Florida poultry farmer adds water to compost to keep the bacteria working that ultimately turns chicken waste into usable topsoil.

Poultry for Profit

Small flocks can produce profit through sales of eggs, meat, and chicks. The niche food market is one of the fastest growing. Consumers are willing to pay a premium for specialty foods.

Joel Salatin's book *Pastured Poultry Profits: Net $25,000 in 6 Months on 20 Acres* is a good resource. Andy Lee and Patricia Foreman's *Day Range Poultry: Every Chicken Owner's Guide to Grazing Gardens and Improving Pastures* is directed at the entrepreneurial poultry owner.

Taking a hobby interest to an income-producing activity is a step that changes the orientation to your avocation. Small flocks offer the possibility of creating self-employment for the determined.

It's not easy, but it is certainly possible. The vendors selling eggs and meat at the local farmers' market are living proof.

107

BREEDING PROGRAMS

Breeding your own birds allows you to perpetuate your flock and select the traits that suit you best, whether that be feather color, body type, or friendliness. You can participate in reclaiming a rare breed. You can join the ranks of knowledgeable insiders at poultry shows.

Poultry keepers have tried every possible way to breed birds over the centuries. The breeding methods included in this chapter are the traditional ones that have satisfied breeders through the years, updated to take advantage of modern tools and knowledge.

As you breed your flock, you will find different characteristics becoming important to you. Each of the following breeding techniques offers advantages. At any given time, you may combine them to achieve the results you want for your flock.

"There is a time to inbreed, a time to line breed and a time to out-cross," said Dick Demasky, a breeder with many years of experience. "Knowledgeable breeders do it all, when the need arises."

DETERMINE YOUR GOALS

Success means different things to different flock owners. As with other livestock, chickens' utility value gets mixed up with their show value. Ideally, small flocks can combine beauty and usefulness. Judges don't evaluate chickens on number of eggs laid or dressed-out carcass weight anymore. Behavioral traits such as broodiness or motherliness are invisible in the show coop. However, if you value these traits, you can cultivate them in your own flock.

Judges do evaluate chickens on body type and feather qualities, including color and condition. The APA *Standard of Perfection* and the *Bantam Standard* are the guides to those qualities. You can acquire those references and use them to help choose which birds to breed. The APA *Standard* is available through its website, $14 for the black-and-white version and $59 for full color. The ABA hardbound *Standard* is $30 through its website; the looseleaf is $12. Some feed and specialty stores may carry these, too.

Those standards are important in maintaining rare and historic breeds. If you decide to dedicate your efforts to conserving a rare or historic breed, breeding to those standards will help your birds keep their historic appearance and qualities.

Raising a flock of a rare breed broadens that breed's genetic base. The more separation there is between flocks of a breed, the greater its chances of survival. Breeding a rare-breed flock for standard qualities and vigor contributes to the overall health of the breed. Even a few generations produces some genetic distance from other populations.

Corallina Breuer

Wing badges make it easy to identify individuals in the flock. They are not permanent, allowing them to be used for a breeding cycle and then removed.

GENERAL POINTERS

Vigor, vitality, and longevity will be significant in your flock regardless of your other goals. A long-lived flock with prolonged utility is always desirable. Selecting birds with long, productive lives will impart low mortality and vigorous constitutions.

Birds selected for breeding should be firm and well muscled without being fat. Legs should be structurally correct. Eyes should be bright, clear, and properly placed. Wings should be carried properly.

Hatching early in the season gives your birds more time to grow during their first year. Your best producers will start laying at a young age, giving you the tip that it's time to start collecting her eggs and setting them to hatch, either under her, under another hen, or in an incubator. Hens that lay well in short-day, natural-light conditions are good producers.

If you know your birds well enough to know them by their eggs or are willing to trap-nest during the entire laying season, select hens that lay through hot and cold weather. The best producers lay through any weather and through their molts.

Color and pattern are important, but some deteriorate with age. A bird with proper color pattern during the first year wouldn't necessarily be culled for deteriorating color as he or she ages, though a bird whose feathers retain the quality of color and pattern over the years would be favored over one whose feathers don't.

Hens producing eggs have large, soft vents. If they are not producing, the vent is small and may be puckered.

Before introducing new birds to your flock, always quarantine them for two to four weeks. Some diseases take time to develop. It's not worth risking your whole flock for a newcomer.

There is no perfect bird. You will always be weighing strong points against weaknesses. That's the challenge of breeding.

TRAP-NESTING

Trap nests are special nest boxes that trap the hen inside until you let her out. The advantage is you can identify which hen laid which egg. The disadvantage is that it's labor intensive: You have to check the boxes at least twice a day to release the hens.

Eggs collected in trap nests like this one can be positively identified as coming from the hen trapped in the nest. This allows the breeder to know what breeding produced the results from that hatch.

Corallina Breuer

Trap-nesting allows you to select eggs from your best birds. Mark them with her number or some other unique identifying mark. You can write on the surface of the egg with a pencil or marking pen. If your breeding records allow, you can identify both parents on that egg.

Trap-nesting also allows you to keep track of egg production. As hens age, they generally lay fewer eggs. If egg production is your goal, you will want to cull those who are not laying to your standard. However, hens with exceptional laying records should be retained as breeders even after their production declines.

Some birds lay odd-shaped or thin-shelled eggs. Using a trap nest, you can identify these hens and cull them. You can also identify such birds by simple observation in small flocks. Soft-shelled eggs are often a temporary condition, so don't cull too soon. A soft-shelled egg sometimes results from two eggs coming down the oviduct in quick succession, leaving inadequate calcium for the shell of the second egg. A hen could be an excellent producer on a double-egg day.

BREEDING PROGRAMS

Beginning breeders may want to start by focusing on utility qualities that are easily measured: the dozen (eggs) and the pound (meat). Appearance seems obvious, but *Standard* qualities can be subtle and elusive. If possible, work with an experienced breeder to develop your eye for the refinements of show qualities. These individuals have spent many happy hours observing their birds. Sitting and watching will educate your eye to the desired qualities.

A single hatch during the breeding season may be enough for you. Two are possible, and some birds are willing to raise three sets of chicks in a year. All chicks from that year are considered the same generation or breeding cycle. A single cycle could extend from January to October.

Different breeding methods require different amounts of record-keeping. Formal pedigrees are not kept in a breeding registry for chickens as they are in many other livestock breeds. Some specialty breed organizations, such as the Serama group, are investigating creating breed registries. Breeders of very rare chickens usually know about the lineage of their birds and their particular breed.

Birds need to be identified, either as individuals or as members of a group. With the advent of the National Animal Identification System, the federal government may require all birds to be marked, whether you plan to breed them or not. Leg bands, wing bands, and toe punches are all good ways to identify your birds.

Leg bands are the standard for identifying birds at poultry shows. Each band is numbered. They are easy to apply and change on a growing bird. They come in all sizes and are color-coded, making it easy to identify them by hatch year or other group characteristic.

Wing bands can be applied to chicks, allowing the feathers to grow over the band. If placed on an adult bird, they interrupt the smooth appearance of the wing. They are more exacting to read, requiring that the bird be caught to read the number.

Toe-punching is done on chicks within a day or two of hatching. You can get a toe punch from the feed store or any poultry supply store. Determine a sequence of holes and notches on the four webs between the toes to provide a unique identification, whether of individuals or groups. The hole or notch usually does not fill in or heal with a scar, so the bird is permanently marked.

Obtain the best stock you can, but don't be afraid to start with imperfect stock. That's part of the challenge. The reward is the gratification of overcoming those imperfections and improving your flock. "The pleasure of surveying

Barry Koffler

Wing bands placed on chicks are later covered by wing growth and are unobtrusive. Individual identification is important in determining which birds to select for breeding.

This adult bird's wing band does not interrupt the feathers on his wing. Wing feathers are important show characteristics, and care must be taken not to damage them.

Barry Koffler

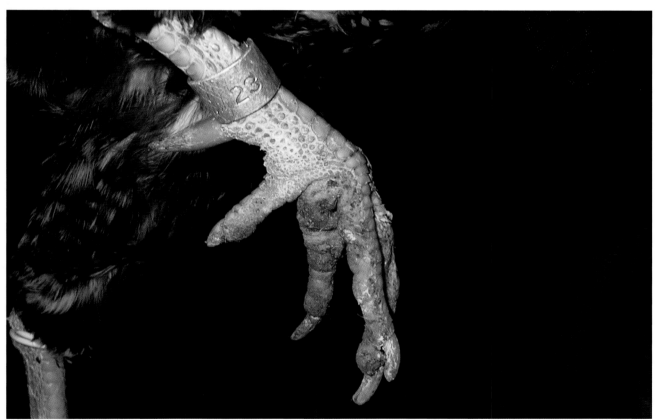

Barry Koffler

Metal leg bands are often used to identify chickens. They are usually permanent and applied when the chicken is blood tested for the first time.

a handsome, uniform flock that has grown from a successful breeding program is hard to beat," says Craig Russell of the SPPA.

ROLLING MATINGS

Rolling matings require the least record-keeping. This breeding system doesn't require any records at all, but it's to your advantage to keep track of the year the birds are hatched, so that you know their age.

You can begin rolling matings with a trio in a single pen the first year. Future years will require additional pens as you separate the best birds for mating.

After the first year of raising chicks, select the best birds and breed them back to the best that you started with. The best pullets go into the breeding pen with the best cock and the best cockerels go into pens with the best hens from the previous year.

At the end of the season, select the best of the old birds and the best of the young birds. They will be your breeding stock for the coming year. That's what keeps the system rolling.

The ratio of the sexes needs to be managed. One rooster to ten hens is about right. Too few hens can result in stressed birds if the rooster is too vigorous. Heavy and feather-footed breeds may do well with only eight hens. Light breeds may do well with a dozen hens.

Birds in large enclosures or on pasture will give each other adequate space. If the hens are being stressed by too much attention from the rooster, take him out every other day or move him from pen to pen.

Rolling matings also allow you to select birds for a particular strength, separate them for breeding, and then work the resulting birds into the main breeding program. These side-matings can help you focus on desirable traits and experiment with breeding in your flock.

Side-matings are any matings done outside your main breeding system. If an unusual characteristic emerges in some of your birds, you can even breed brother to sister, although that has the highest risk of bringing out undesirable traits. You can then include the best individuals in your regular breeding program.

You will need at least two pens for each breed or variety. Additional pens may be needed as your flock expands.

LINE BREEDING

Line breeding results in daughters being bred back to fathers and grandfathers, sons back to mothers and grandmothers. It is a modern method that has worked well to produce rapid gains in characteristics. It is used to perpetuate strong qualities or individuals. Line breeding results in predictable and reliable progeny. Surprises are unlikely, because the genetics are limited.

Weaknesses can be identified and those birds culled. Keeping a generation between breeding partners reduces the risk of inbreeding weaknesses.

The best birds are considered "seed stock" and are used only for breeding.

Line breeding differs from rolling matings in that it breeds progeny back to direct relatives. Rolling matings involve chickens from various matings bred to the best roosters and best hens. In the example of having only one trio, this would be the same as line breeding, but in most cases breeders will have many more chickens and they all switch around.

CLAN MATINGS

Separating your flock into clans, or yards, along either matriarchal or patriarchal lines, allows you to maintain vigor in the flock without introducing new birds. Demasky has maintained a flock of Old English Games since 1966 using this method.

In matriarchal clans, each hatching egg is marked with the mother's clan mark when it is collected. As they hatch, each chick is toe-punched and wing-banded with the clan mark. It's usually convenient to hatch batches of each clan together. Additional marks can also identify individuals and link them to their ancestors.

Chicks can be kept separate by hatching all the eggs of a single breeding under one hen or in one incubator or on a particular day. Small wire hatching cages can keep the chicks separate as they hatch. In patriarchal clans, all eggs from the hens bred by that rooster are marked and incubated together.

Clans are maintained as separate stock. Cocks and hens of the same clan are never bred to each other. They are always bred to birds from other clans, so you need to keep at least three clans. Breeders typically keep an odd number of clans.

Matriarchal clans are usually pair-mated. A particularly successful pair can be kept together for years. You can experiment with other pairs for different results.

In patriarchal clans, you set the rooster in with a group of hens. A rooster can be bred to any hens not in his clan.

The variety of possible matings reduces the possibility of depending excessively on any individual bird. An especially desirable individual's qualities can be perpetuated by creating clans of full sisters or brothers.

Clan matings require keeping records of each chick. In matriarchal clans, it's possible to trace the ancestry of every chick. These pedigree records can give you a rich store of detail about your birds and your breeding system.

BREEDING OUT-AND-OUT

Breeding out-and-out means introducing new roosters every one or more years. This brings new genetics into your flock, increasing diversity.

Corallina Breuer

This Sumatra rooster and several hens are well matched. Their long black feathers have a green shimmer; their long, sweeping sickle and covert feathers give them a graceful carriage; and they have almost no wattles.

Corallina Breuer

Silkie hens are known for their broodiness. Broody hens stop laying eggs, so egg producers consider this trait a fault.

You may keep records on each rooster. With some very rare breeds, only a few lines exist. Even if you acquire a rooster from a different person or hatchery, he may be from the same genetic line as your current birds. Keeping records helps you identify your stock and its relationship to other flocks.

This method can produce unexpected results, reducing uniformity in your flock. It will keep your flock productive and may offer you characteristics you want to perpetuate in side-matings.

GRADING

Grading introduces stock from another breed to the existing flock to add a characteristic or change it in some way. Grading is often used with rare breeds that have lost genetic vigor. By adding birds of another breed, then breeding the offspring back to pure birds, flocks can be reinvigorated.

Grading can be controversial because it adds different genetics to the purity of the flock. But that purity is relative and can be regained through breeding back to the original stock. The following table shows the progression back to purity over a number of breeding cycles:

Generation	Fraction	Percent
1	1/2	50%
2	3/4	75%
3	7/8	87.5%
4	15/16	93.75%
5	31/32	96.875%
6	63/64	98.4375%
7	127/128	99.21875%
8	255/256	99.615375%

This is an old system that has long been used to maintain other livestock, including cattle, sheep, horses, swine, goats, and dogs. For all practical purposes, eight breeding cycles result in pure stock.

Each bird has to be identified by its group in order to choose appropriate mates and be certain where the flock is in progress toward purity.

This method can also be used to develop new breeds or re-create breeds.

Corallina Breuer

This White Silkie rooster is a vibrant bird for breeding. Heavily feathered breeds may need to be artificially inseminated, but Silkies usually manage natural mating.

Within a breed, this method is called stud mating. One or more roosters with particularly desirable characteristics can be graded into a flock. After breeding them to a group of hens, the sons become the males for the rest of the flock. Only the best birds are bred. You might develop your own line of superior males this way.

CROSSBREEDING

Crossbreeding—breeding a hen of one breed to a rooster of another—can produce hybrids with desirable utility characteristics. The offspring will not be acceptable for showing. The first generation of two distinct breeds bred to each other shows the most hybrid vigor.

Most commercial chickens are hybrid crosses, Cornish on Plymouth Rocks. Commercial breeders maintain separate breeding stock to produce their birds.

Crossing breeds is tempting and has always been done by breeders seeking something new. Popular crosses in the past include Games and Dorkings, usually a Game male on a Dorking female. At the turn of the twentieth century, this cross was thought to produce the best meat chicken.

A Malay male on a Dorking hen produces offspring that grow faster than either parent. A Dorking male on a Brahma female results in a slower-growing hybrid that gets nearly as big as a Brahma and has the meat quality of a Dorking.

SEX LINKS

Birds have separate sex chromosomes just as mammals do. Traits can be expressed differently in the sexes, depending on how a single gene is expressed in males and females. The advantage to sex links is that the males can be separated from the females very early on. Any breeding program needs fewer roosters than hens, so the earlier they can be identified, the better.

The Delaware breed was originally developed as a commercial sex-linked cross, producing males for broilers and fryers and females to lay brown eggs. They have since become a breed recognized by the *Standard*.

Another common cross is between red fathers and white mothers, resulting in red daughters and white sons.

In barred breeds, hens require a single gene for barred plumage; roosters require two. Breeding a solid cock to a barred hen results in solid females and barred males.

The gene for barred feathers is a modifier gene, interrupting otherwise solid feather color. Barred Rocks are usually black and white, but the gene can be introduced into other colors. Buff Barred Rocks exist.

Hybrid Vigor

Hybrid vigor, or heterosis, is the phenomenon of increased strength, intelligence, and other desirable characteristics in the offspring of crossing two pure breeds. In chickens, results vary depending on which breed was the mother and which the father. All crosses are not equal.

That's the basic idea of outcrossing, introducing birds of a different breed into a breeding plan.

If you breed first-generation hybrids to each other, you will not get offspring like the parents. Breeding programs can work with this approach, but don't think that you can breed a rooster of one breed to a hen of another breed and get a flock of a new breed. Breeds are defined by breeding true, which means reliably hatching similar offspring.

The first hybrid generation is usually the one that exhibits the most hybrid vigor.

Ed Hart

Sex-linked chicks show different plumage by sex from the start, an advantage in differentiating females from males.

INCUBATION
AND
CARE OF CHICKS

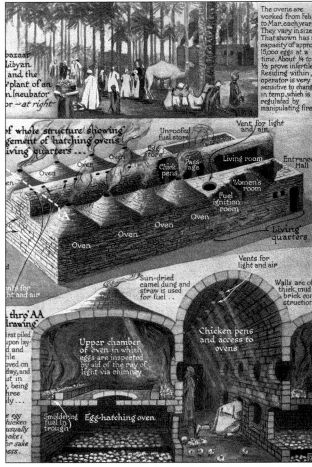

The ovens are worked from Feb. to Mar. each year. They vary in size. That shown has a capacity of appro 15,000 eggs at a time. About ¼ to ⅓ prove infertile Residing within, operator is very sensitive to chang in temp. which is regulated by manipulating fire

bazaar Libyan and the plant of an n Incubator or ~at right·

f whole structure showing gement of hatching ovens living quarters ...

Unroofed fuel store

Vent for light and air

Egg store

Oven Oven Chick pens Pass- age Living room Entrance Hall

n Oven Women's room

Fuel ignition room

Oven Oven

Oven

Living quarters

Vents for light and air

nts for ht and air

thro' AA drawing

Sun-dried camel dung and straw is used for fuel ..

Walls are of thick mud & brick con struction

rst piled upon lay- d and ile. ved on day, and it in , being hree ly ..

e egg hicken usually oke: or sake ess..

Upper chamber of oven in which eggs are inspected by aid of the ray of light via chimney

Chicken pens and access to ovens

Smoldering fuel in trough Egg-hatching oven.

Originally printed in *National Geographic*, April 1927

Hatching ovens in Egypt produced 15 million to 20 million chicks annually. The operator usually lived in the incubator and judged by experience when the fires needed attention.

Natural incubation is the traditional way to keep a backyard flock going. You will need one or more hens who are willing to set the twenty-one days required to hatch chicken eggs and a place for her to do it. Hens can be remarkably resourceful in making their own place to hatch a brood. More than one small-flock keeper can tell a story about the hen they thought was lost, only to have her reappear surrounded by chicks.

Artificial incubation has been around for thousands of years. The Romans figured out how to do it and ancient Egyptians made a business of it. It's a good idea to have an artificial plan in reserve, in the event anything happens to your broody hen. One breeder finished hatching eggs in an electric skillet after a predator killed his broody hen.

Both methods work, although the demands of artificial incubation may give you new respect for your chickens. Once artificially incubated chicks have hatched, you will need to take special care of them unless you have a hen that is willing and able to care for them.

BROODINESS

Hens may signal their intention to become broody by setting on the nest and refusing to move. You can stimulate them to become broody, for your convenience and timing to the eggs you are collecting, by starting them on a nest of artificial nest

eggs made of wood, plastic, or glass. Some hens move on and off the nest for a few days before getting serious. Give her time to get used to the idea. When she stays on the nest for at least 24 hours, she's ready to get serious about it.

A clutch is the group of eggs hens want to collect before beginning to set. One of the marvelous things about hatching eggs is that although an individual hen will lay one egg a day, she will keep on laying until she gets a clutch before beginning to incubate them. Then they all hatch together.

Corallina Breuer

This hen concentrates on laying her egg. Hens often prefer to lay in nests with other eggs, but some will lay anywhere.

Corallina Breuer

Hens often look for eggs in a nest when choosing a place to lay that day. Hens sometimes crowd on top of each other in a nest box or line up waiting for their turn. When a hen becomes broody, she may even peck at other hens who try to enter the nest to lay their eggs.

This Silver Laced Wyandotte eyes the brown egg she has just laid. Hens are usually excited and proud about laying an egg, often announcing the event with loud clucking.

Old Incubators

Don Sherrick of Sunny Creek Farms in Red Lake Falls, Minnesota, has made incubators his lifework. He turned seventy-five in 2006 and had been in business for fifty-nine years.

"I originally started out with geese and got addicted," he says. "Rigor mortis is the only thing that cures this addiction."

Many old incubators remain unused in farmhouse basements. Sherrick is always on the alert for a find. Incubators were valued by their owners and often remain in excellent condition. "Anybody who has never used a redwood incubator doesn't appreciate it," Sherrick says.

He can supply manuals for most old incubators. If he doesn't have a manual, he can tell you about how to work it. George Koke of Franksville, Wisconsin, found a 1940s Jamesway 2940K incubator with a hand-crank turner and needed some help getting it going. Now, it's hatching around 90 percent of the pheasant eggs he incubates.

"Now I have to figure out how to feed all these birds I'm hatching because of this incubator," Koke says.

Sherrick's first incubators were kerosene-fired, sending carbon into the air and blackening his mother's curtains if he didn't keep the wicks trimmed.

"My parents were never bird-oriented," he says. "They were practical people."

Practically enough, Sherrick has made a successful business of restoring old incubators and selling new ones on his farm. He sold one to the Flamingo Hotel in Las Vegas to hatch the penguins they exhibit.

Sherrick and his wife, who is a registered nurse and not a bird person, raised four children on their farm.

"A lot of people said, you'll never make a business of it, but I looked at it as a business from the start," he says.

Broodiness

Broodiness is the behavior a hen exhibits when she is ready to set on eggs until they hatch. She will stop laying eggs and take over one of your nest boxes.

The interruption in egg production is the reason this trait has been bred out of may breeds. Perpetuating your own flock is a reason for nurturing this characteristic.

Hens, like other birds, look for some number of eggs that signals a clutch is ready to be hatched. Usually, it's eight to twelve eggs. That's the reason hens often take turns laying in a single nest. They are looking for that magic number.

Corallina Breuer

This Blue Cochin hen is a good mother to her new chicks. She will teach them how to forage for food and find their places in the pecking order.

BROODY BREEDS

Not all breeds are created equal when it comes to broodiness. In many breeds, broodiness has been actively selected out, since broody birds stop laying eggs. Today's Dominiques do not have the same level of broodiness as their forbears, ever since an attempt in the 1920s to turn them into a thoroughly modern breed.

Heritage breeds that brood well include Dorkings, Games, Javas, Plymouth Rocks, Wyandottes, Nankins, Dutch, Brahmas, and Orpingtons. Silkies are particularly well known for their broodiness and good mothering, as are Jungle Fowl. Oriental breeds usually make good brooders and mothers, including Phoenix, Yokohama, and Cubalaya breeds. In general, show strains retain their broodiness and commercial strains do not. Bantams are more likely to be broody than large fowl.

There are other factors to consider as well. Game breeds have hard feathers, with narrow, short shafts and closely-knit barbs, making the feather stiff and shiny rather than fluffy. They cannot cover as many eggs as a fluffy breed such as the Dorking or the Brahma.

HATCHING

A separate chicken house is desirable for hatching. The hen should be undisturbed and undistracted by her sister hens. The house should be kept clean with fresh food and water always available.

A broody hen will get up once a day to eat, drink, and defecate. You may notice large piles of feces, because she only evacuates once a day. Keep the house clean and quiet, dark but not entirely without light, so she can see what she's doing. The eggs will cool off slightly when the hen is off the nest, which is a normal part of their cycle.

Provide a nest similar to the ones in which the hen is accustomed to laying her eggs. Add enough straw or other nesting material for her comfort and to cushion the eggs. Size the nest box to the hen. Large hens may be ungainly in nests that are too small, resulting in broken eggs.

Broody breeds

Although hens of all these breeds are likely to be good brooders and good mothers, they vary by individuals as well. All games are usually good brooders.

Ameraucana	Dorking	Old English Game
Asil	Dutch	Orloff
Barnevelder	Faverolle	Orpington
Brahma	Holland	Polish
Buckeye	Japanese	Plymouth Rock
Chantecler	Java	Rhode Island Red
Cochin	Jersey Giant	Silkie
Cornish	Kraienkoppe	Sussex
Cubalaya	Marans	Welsummer
Delaware	Nankin	Wyandotte
Dominique	New Hampshire	

Hens may also be set directly on the ground inside a coop. Hollow a spot out of the ground and line it with hay or straw. Add a few artificial nest eggs to attract her interest. Place the hen in it. Allow her time to arrange the nest to her liking before adding the hatching eggs.

Eggs can be set under any capable hen. You can collect eggs from other hens or even put eggs of other bird species under her. Large eggs need large hens to cover them adequately, but a small hen with good intentions can manage a small clutch of large eggs.

SELECTING EGGS TO HATCH

The eggs you collect will be determined by your breeding program. They may include the brood hen's eggs or not.

Not all eggs are good candidates for hatching. The best-shaped eggs have a large end and a small end. Some hens regularly lay double-yolked eggs. Although these are often popular with egg customers, they will not hatch. Avoid all odd-shaped eggs. Eggs should also have the correct color for their breed.

Inspect eggs for cracks and other imperfections. Hold the eggs to the light to examine them for checks, thin spots in the shell that do not penetrate the shell. Eggs with imperfect shells are not good candidates for hatching.

Severely stained eggs and eggs with manure crusted on them should not be hatched. The average eggshell has 17,000 pores through which air passes. Bacteria can enter and contaminate the egg, destroying the embryo.

Eggs are seldom perfectly clean, however. They can be washed in a diluted bleach solution; in Oxine, a popular disinfectant; or in plain warm water. Cold water can cause the eggshell pores to contract, drawing bacteria into the egg.

Eggs can be collected for a week or more before placing them under the hen. Hatchability declines severely in eggs stored longer than seven days. Embryos will begin to develop after the eggs are placed under the hen, so they will all hatch together. Keep the eggs cool, approximately 55 to 60 degrees Fahrenheit, and undisturbed until you are ready to set them, but do not refrigerate them. A basement is a good place to keep eggs until you are ready to hatch them.

Eggs can also be shipped. Eggs you receive should be allowed to rest overnight before setting them under a hen.

Because shipped eggs have been subjected to jostling along the way, hatching rates are generally lower than for eggs that come from your own hens. Any extra disturbance can interrupt development of the embryo, so always handle hatching eggs gently.

Corallina Breuer

These dark brown Welsummer eggs are being incubated in a round incubator. The thermometer on the side tracks temperature. The thermometer should be placed vertically one-quarter to one-half inch below the top of the incubating eggs, to reflect the temperature at the level of the developing embryo.

ARTIFICIAL INCUBATORS

Artificial electric incubators can do a good job. Small ones that hold less than a hundred eggs must be monitored for temperature and humidity; large modern ones that can hold hundreds or thousands adjust these conditions automatically. Older incubators do not always have this feature, but breeders who use them learn to adjust those conditions themselves. Many advocate for their equipment.

"Often, the seventy-year-old machines have better hatch rates than some of the newer models," says Monte Bowen, who uses a 1930s Sears Roebuck Farm Master 400 to hatch his Javas, Black Dutch bantams, and Nankin bantams. Bowen credits the Farm Master 400 with a 98 to 99 percent hatch rate.

Chicken eggs need a constant temperature of 99.5 degrees Fahrenheit. Variations in either direction will affect the hatch rate. The eggs benefit from twenty minutes of cooling each day. This corresponds to the time the hen would spend off the nest eating and drinking.

The air in the incubator must be kept humid. Humidity pads provide constant water to the warm air. A wet-bulb thermometer can allow you to calculate relative humidity inside the incubator. Follow the manufacturer's directions.

The incubator must have proper ventilation to provide fresh air to the developing embryos, which give off carbon dioxide. The warm air will circulate on its own in small incubators. Larger ones should have a fan or some other way to assure the air circulates.

Transfer the eggs into a hatching tray on the eighteenth day, three days before you expect them to hatch. A hatching tray is covered or otherwise enclosed so that the chicks can't get out. The humidity should be increased from 50 to 55 percent to 65 percent. The temperature can be left at 99.5 to 100 degrees until the chicks hatch. Then it can be lowered to 95 degrees.

Consider the possibility of a power outage. If possible, have backup power available from a generator. If no backup power is available and the power is interrupted, don't despair. Leave the incubator closed to maintain the warm air as long as possible. A few hours of cooling may delay the hatch a bit, but it isn't necessarily fatal to the entire hatch.

Think of that hen and her patient faith. Give it time.

Vintage Incubators

Jamesway, Petersime, Robbins, and Humidaire are twentieth-century incubator brands that you are likely to encounter today. Although they are no longer in production, these incubators are so sturdy and successful that many breeders still swear by them.

Jamesway 2700, 2940, and 252 from the 1930s, 1940s, and 1950s are all popular electric incubators. They were preceded by coal-fired and kerosene incubators that operated on hot-water convection systems. They were designed to fit through the doorway of the farmhouse for small-flock operations.

Petersime proclaimed itself the biggest incubator factory in the world back in the 1950s. Its design placed trays of eggs in a central drum and circulated air by a paddle system. The company closed its doors in 2005, but devoted breeders still use the incubators.

Redwood was the best material for incubators. Robbins built its incubators out of this type of wood for more than thirty years. Green fiberglass in the 1960s was not an improvement, and another change to insulating material sandwiched between light metal sheeting in the 1970s did not save the company.

The Museum of Science and Industry in Chicago has used a Humidaire incubator since it bought its first one in 1949. After an unsuccessful attempt using a modern brand, the museum sought a replacement of its original.

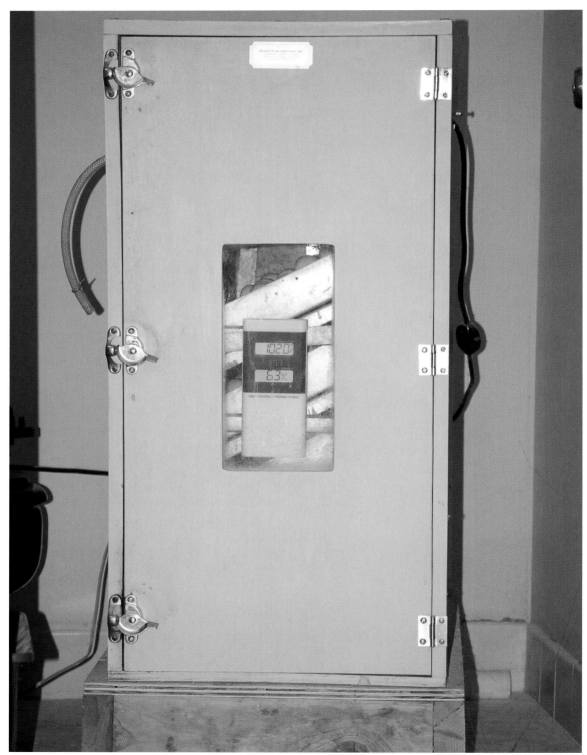

Corallina Breuer

This large modern incubator at the University of Wisconsin holds hundreds of eggs in trays that shift from side to side to accomplish turning. The full-sized door gives easy access to eggs.

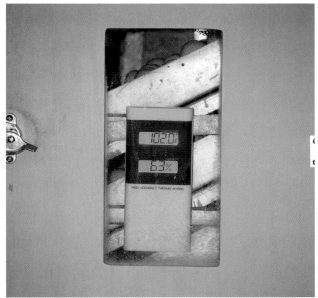

Corallina Breuer

Digital readouts on the door monitor temperature and humidity. Eggs need to incubate at 99.5 to 100 degrees Fahrenheit for forced air, 102 degrees for still air. Humidity needs to be kept at 56 to 60 percent.

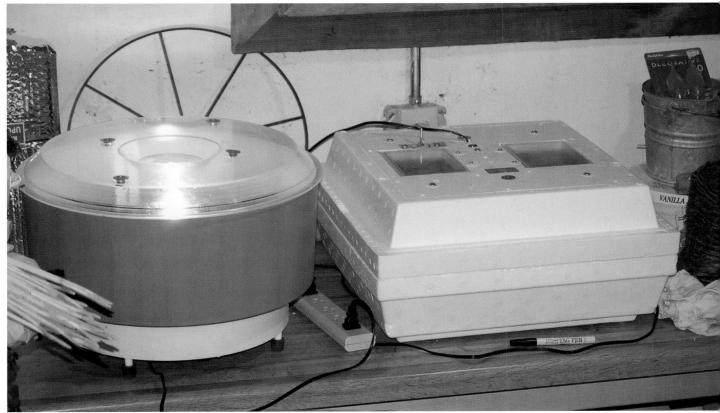

Corallina Breuer

Small tabletop incubators like these are adequate for backyard flocks. They hold a dozen or more eggs. Eggs may need to be turned by hand. A written schedule and pencil marks on the eggs help keep track.

TURNING EGGS

A broody hen will turn her eggs many times a day. She is businesslike, carefully maneuvering each egg to get it into a position that pleases her.

Large artificial incubators turn the eggs automatically. The eggs fit into the trays with the large end up.

If you are using an incubator that doesn't offer that convenience, you will need to do it yourself. Let the embryos get established, undisturbed, for the first three days, then start turning them on the fourth day. Turn them at least three times a day until day 17. Do not turn eggs after day 17. The embryos are moving into position for hatching.

Post a paper near the incubator and record when you turn the eggs to keep track of when the eggs were last turned. Chicken eggs do not need to be completely turned over. Rolling your hand across the tray of eggs is adequate.

CANDLING EGGS

You should candle eggs on day 10 and day 18 of incubation. A candler is any bright light source that can be directed toward an egg. A strong flashlight with a narrow beam or a toilet paper tube will do. You can cut a one-inch hole in a piece of cardboard to limit the light source.

You will be able to see shadows inside the egg if it is developing. If not, it will be clear. Developing embryos may be seen as a "spider" of veins as early as day three. Embryos may start developing and then stop; these are called "quitters." You will be able to tell the difference between quitters and the eggs that continue to develop. Eggs that never start to develop should be removed. Quitters should be removed, as they may explode and make a terrible mess.

You should see the embryo developing by day 10. Day 18 is just three days before hatch, so the egg should be completely dark except for the air sac, which should be clear. You will see no movement.

HATCH RATE AND FERTILITY RATE

The hatch rate is the percentage of eggs that hatched out of the eggs that were originally set. If you set ten eggs under a hen and nine hatch, the hatch rate is 90 percent.

Many factors influence the hatch rate, beyond how good the hen was. The eggs may have been chilled or shaken, dooming them before they started.

Fertility rate is the number of eggs that begin to develop, whether they arrive at a live hatched chick or not. Fertility does not have to be 100 percent to be good. Duds are inevitable.

However, fertility below 80 percent indicates some problem in the flock. Are you sure the rooster is fertile? Does he have adequate access to the hens? Are there too many hens for him to do justice to?

Viability of fertile eggs is the percentage of live chicks from fertile eggs. Fertile eggs may not proceed to develop into mature chicks for many reasons. Weak embryos, which begin to develop but then "go stale" and die, can be the result of genetic weakness in the parent stock, poor humidity control, or temperatures that are too hot or too cold.

Keeping records of all these results of your hatching seasons will help you detect problems. Low fertility and inept brooding will influence your choices of chickens for future breeding programs.

CARING FOR CHICKS

Chicks hatch by pecking a circle around the pointed end of the egg with their egg tooth. It takes anywhere from one hour to several hours for them to get out of the shell. They peep before they hatch. Give them time.

It's tempting to help the chicks get out, but most experienced breeders agree that letting them succeed on their own is an essential part of maturation. Chicks unable to overcome this first obstacle may have other genetic weaknesses that make them undesirable.

Breeders may make an exception in the case of a rare-breed egg. If individuals are so scarce that each one represents an important contribution to the flock, helping it out of the egg is acceptable. Its other genetic weaknesses may be compensated by other breeding tactics.

For the first month of life, chicks require special living conditions provided by a brooder. In the case of naturally hatched eggs, that's the mother hen. Otherwise, a heated, enclosed brooder can be purchased or homemade.

Chicks need to be protected from drafts and kept warm, while allowing for air circulation. The brooder needs

Corallina Breuer

White Silkie, Black Cochin, and Blue Cochin chicks are kept in the brooder for a day. Humidity should be increased to 65 percent during hatching. They need no food or water for the first 72 hours, as they continue to draw on nourishment from the yolk.

These Cornish/Rock crosses, the chickens raised for commercial meat production, are getting their first feathers. Chicks huddle together for warmth. If they are bunched up in a pile under the heat source, they are not warm enough.

Corallina Breuer

Corallina Breuer

White Silkie chicks start out with a crest. Black Cochins and Blue Cochins are fluffy, as they will be as adults. The temperature should be decreased about five degrees a week after hatching, until chicks have enough feathers and maturity to manage ambient temperature.

Corallina Breuer

Some differences show up in chicks from the start. This Standard Blue Ameraucana chick shows its light coloring from the start. Sexes are not so easily distinguished in this breed.

sturdy sides. Wood and cardboard are good materials. Some poultry keepers use children's wading pools, which can be covered and enclosed on one end and screened on the other. A heat source at one end—either a regular light bulb or heat lamp—and food and water at the other end ensures that the chicks will get some exercise running back and forth.

The heat source should cover about a quarter of the brooder and be placed close to a corner, but not so close that the chicks pile on top of each other and smother one another. You can use a thermometer to monitor temperature; start at 95 degrees for the first week and reduce it 5 degrees each week.

The chicks will be the best guide to temperature. If they lie down in a wide, open circle, it's too hot and the heat source is too close to the floor. If they huddle together under the light and cheep loudly, it's not warm enough. If they are sleeping close together and peeping contentedly, they are just right.

Chicks grow best on light metal mesh that allows their droppings to fall through for easy cleaning and gives their feet something to grab on to. Chicks raised on newspaper risk deformed legs and feet.

For information about vaccinating against diseases like Marek's disease, see Chapter 12: Health Care.

FOOD AND WATER

Chicks don't require food or water right after they hatch, because they are still absorbing yolk, but both should be available right away. Put a line of feed on a piece of newspaper at the open end of the brooder. When this is messed up, you will know that your chicks are ready to eat.

Chick starter ration is about 20 percent protein. Keep the feeder full so that small chicks can reach it. Feed them chick starter for the first ten weeks, then switch them to a mix of four parts scratch to one part starter until they are sixteen weeks old. Larger breeds, especially those of Oriental heritage, are inclined to develop leg and heart problems if they grow too fast.

Make sure the waterer is safe. A pint waterer is small enough that even bantam chicks can't drown, whereas a larger waterer may be risky for them. Put marbles in the trough for a week or more until the chicks are large enough to get out if they fall in and can drink without getting too wet.

Robert Plamondon's *Success with Baby Chicks* (2003) has more details on giving your chicks the best start in life.

ENJOY THE REWARDS

Newly hatched chicks are truly a miracle of nature. Take time to appreciate them.

Successful hatching is an inspiring part of breeding your own chickens. It also gives you bragging rights among your chicken-owning colleagues.

SELECTION
AND
CULLING

Selective breeding is the central issue in developing breeds. In the wild, environmental factors make the difference between which individuals survive to maturity and breed, passing on their genes and characteristics to the next generation. That's natural selection.

Artificial selection occurs when humans get involved. When they decide one individual is more desirable than another, they manipulate breeding to enhance some traits more than others. The basic biology is the same in both cases and hasn't changed since humans started domesticating animals.

The process is rarely simple and straightforward. With uncounted numbers of genes and infinite possible combinations, teasing out the desired traits from the unwanted traits is a delicate process. All characteristics are governed by at least two genes, some of which are expressed when only one gene is present and some of which require both genes for that trait to be expressed. Genes may mix and combine to result in different colors or other characteristics.

Breeding chickens is a life lesson in starting where you are. The stock available to you may be peppered with faults and defects, but it is what you have. Appreciate their strong qualities and build on them.

Improving a flock by breeding to the *Standard of Perfection* requires time and patience. The rewards of seeing a healthy flock of admirable birds that you have influenced toward perfection are well worth the effort.

Humans have been influencing chicken breeding since before recorded history. You and your breeding flock join a long line of chickens and breeders.

GENERAL PRINCIPLES

Generally, two genes (genotype) govern the expression (phenotype) of every trait. Some genes require that both be the same for the trait to be expressed. These are recessive genes. If a recessive gene is paired with a dominant gene, the dominant trait is expressed.

In some breeds, the single-comb trait is recessive. The rose-comb trait is dominant. If you produce rose-combed birds in breeds that are required by the *Standard* to have single combs, you will have to breed single-combed birds into the flock until your birds are homozygous, meaning both genes are the same, and have single combs.

Genes don't always act in clear-cut ways. The rose-comb gene dominates over the single-comb gene, but it interacts with the pea-comb gene to produce an intermediate comb called the walnut comb.

Several kinds of genes govern white color. Some are dominant and some are recessive. Breeding Barred Rocks with entirely colored New Hampshires produces black chicks. Go figure.

Breeders before you have worked to discover how genes influence traits, although many questions remain unanswered. Explore published information and the Internet for details on the genetics of your chosen breed as you make decisions on your breeding program.

INBREEDING

The desirable traits you want to perpetuate in your flock unavoidably come with all the other traits. As you breed your flock toward certain goals, the genetic variability decreases. A small flock with limited numbers of individuals is inevitably restricted in its genetics. Breeding closely related individuals isn't necessarily bad, but it does increase the possibility of enhancing weak or undesirable traits along with the ones you want.

Breeding genetically similar individuals results in a phenomenon called inbreeding depression, a loss of vigor or fitness. It happens when genes pair up and create undesirable or even deleterious traits that are masked by dominant traits. The resulting individuals may be weak in general ways, such as reduced hatchability or chick mortality, or have undesirable characteristics such as wry tail, in which the tail is permanently bent to one side.

Inbreeding problems are avoided by keeping a large flock with more than one rooster and using other breeding strategies discussed in Chapter 8: Breeding Programs.

Breeding to genetically dissimilar birds restores vigor but may also dilute the qualities you seek. Consider the whole bird and how a potential breeding will change its progeny in all characteristics, not only the ones that interest you most.

Rarity confers special value on individuals that would not normally be considered desirable. Stock does not need to be perfect. Selective breeding allows you to influence the development of your flock and move toward perfection.

GENETIC DRIFT

Genetic drift is the reduction in genetic diversity that results from breeding a small number of individuals. One individual's characteristics can become more pronounced, a phenomenon called founder effect. A flock drastically reduced, perhaps by a predator attack, may experience a bottleneck, or a reduction in genetic variability.

Corallina Breuer

This closed flock of White Plymouth Rocks has been maintained at the University of Wisconsin for about twenty years without loss of vigor. Careful attention to breeding can avoid the pitfalls of small gene pools.

Characteristics can be fixed in a small flock or can drop out and be lost. The effects of genetic drift are random rather than deliberate. Unlike inbreeding, it doesn't necessarily mean that the flock is weaker or less vigorous.

Less genetic diversity does mean the flock has fewer characteristics to allow it to adapt to new circumstances. Rare breeds may face this problem.

Flocks affected by genetic drift can be reinvigorated by the same breeding techniques used to avoid inbreeding. Introducing additional hens with the desired characteristics to the flock will increase genetic diversity. Maintaining small flocks separate from each other and then periodically crossing them with each other can help. The more small flocks of a rare breed being raised in geographically distinct parts of the country, the better chance the breed has of surviving.

ARTIFICIAL SELECTION

Selective breeding is consciously choosing individual birds with certain characteristics to breed. Whether that be for comb type, back length, feather color, or some other characteristic, each comes with the rest of the chicken. Characteristics that are not desirable may be perpetuated along with the desired ones. Groups of genes that work well together (known as co-adapted gene complexes) may be inadvertently broken up.

If you are breeding toward a goal, the more detail you can keep on each breeding cycle, the better you will understand how your breedings are working out.

Breeding birds should always be healthy and vigorous. Sickly birds may pass along constitutional weakness.

SELECTING FOR PRODUCTION

Production goals include feed-to-meat conversion and egg production. Know where you are starting: what your feed costs are and what weight the birds reach by the time you are ready to process them. Egg production can be determined by trap-nesting the hens for a period of time.

Chickens that have the highest feed-to-meat conversion and lay the most eggs become your breeding birds. They will then leave the production flock, becoming the founders of the breeding flock.

Thoughtful breeding mindful of the pitfalls of inbreeding and genetic drift will enable a small-flock owner

Corallina Breuer

This hen is laying her egg in a trap nest. She springs the trap when she enters and cannot leave until the keeper releases her. It's a way of identifying which hen laid which egg.

to perpetuate a production flock without loss of genetic diversity or healthy vigor.

SELECTING FOR CONSERVATION AND SHOWING

Rare-breed conservation requires different strategies. The *Standard of Perfection* describes the physical characteristics required for each breed and variety. A breed's history can inform you of typical behavioral characteristics that are often genetically influenced.

Veteran breeders advise, "Build the barn before you paint it!" In other words, the bird's type is more important than its color. The size and shape of the body define the breed's type. Games are long-legged, Javas have a broad back.

Meat- and egg-production values are an integral part of traditional breeds. Type does not conflict with good production.

This sounds easier than it is. Variations on type may be subtle to the untrained eye. Your sensitivity to variations in your breed of choice will develop as you attend shows and observe your breed closely. The advice of experienced breeders is invaluable.

Traditional breeds may take a year to reach maturity and reveal their fully developed body type and feathering. They need time to grow. Except for obvious deformities and disqualifications, avoid culling too early or you could deprive yourself of a significant bird.

Older birds may exhibit desirable changes in feather pattern. Mottled Javas that have too much black in their first year may have perfect color patterns in their second or third year, as white increases with each molt. An individual with enough white in its first year may have too much white in succeeding years.

Other feather colors may fade with age. This characteristic is not a disqualification, although of course birds that retain full color are preferred.

DEFECTS AND DISQUALIFICATIONS

Monte Bowen of Kansas, an experienced breeder of Black Dutch and Nankin bantams and large-size Javas, is known for his insistence that fanciers "breed to the *Standard*!" The APA and ABA *Standards* define the parameters against which birds will be judged at shows. They have been thoughtfully developed over time and are not changed without careful consideration and much debate.

No bird is perfect. All fall short of the *Standard* in some respect. Minor failures to meet the *Standard* are considered defects. Defects include being too large or too small, comb irregularities, unsatisfactory feather quality, underdeveloped tail or wings, and incorrect feather color. Defects cause points to be subtracted from a bird's score at the show, but they may not require that a bird be excluded from breeding.

Serious discrepancies from the *Standard* are disqualifications. They are so significant that birds entered in shows will be disqualified from judging. Such problems are often genetic. Birds with disqualifications should be culled and not included in breeding programs. Disqualifications include being 20 percent or more overweight or underweight for the breed, misshapen combs, deformed beaks, blindness or other eye irregularities, twisted wing feathers, split wings, crooked backs, split tails, wry tails, bow legs,

Corallina Breuer

Traits like this crossed beak are never acceptable. This bird would never be used in a breeding program. With care to make sure the bird gets enough to eat, hens with crossed beaks can be good laying birds or affectionate companions.

deformed feet, knock-knees, webbed feet, the wrong number of toes, the wrong color of face or ear lobes, and some feather colors on certain breeds. Split wings have permanent separation between the primary and secondary feathers due to the absence of one or more feathers; split tails have gaps in the tail feathers due to the absence of feathers or improper placement of tail feathers.

SELECTING BREEDING BIRDS

Since no bird is perfect and all have some faults, the challenge of the breeder is to approach perfection while minimizing faults. You will always be breeding less-than-perfect birds to each other.

Never breed birds that share the same fault. It will inevitably appear in the offspring and may become fixed in your flock. Eventually, you would have to breed it out.

Evaluate the strengths and weaknesses of your flock. Determine which traits you want to preserve, which ones you want to change, and which ones need to be eliminated. Each one will carry some weight in making decisions on which birds to breed to which.

BEHAVIORAL TRAITS

Broodiness and other traits of disposition and behavior are less obvious but no less important to the health of your flock or your enjoyment of it. Friendliness, fertility, egg-laying, and adaptability to local climate may also go into your list of qualities to perpetuate in your flock.

Broodiness has been bred out of many traditional breeds. Natural incubation is the best choice for small sustainable flocks, and there are few joys to compare with watching a hen care for her chicks.

Attacking humans, otherwise known as man-fighting, is not acceptable. Breeding a chicken with this quality will only bring you grief, no matter how desirable the bird's physical characteristics. A rooster should be protective but not aggressive.

Fertile eggs are crucial to successful breeding programs as well as to the perpetuation of rare breeds. This quality often loses ground due to inbreeding in breeds that have small populations. The individuals who show the highest rates of fertility and hatchability are preferred to individuals who may have strong physical traits but rank poorly in passing them on.

Egg-laying may or may not be important to you, but it has always been one of the main reasons for keeping chickens. Traditional breeds would not exist if they were not productive. The biological and behavioral basics required to reproduce secure the survival qualities that all species need to maintain their existence.

Most breeds do well across a wide variety of climatic conditions, so adaptability may not be important in your geographic area. However, marshy conditions, hot and humid weather, desert conditions, and very cold climates will stress birds. Choose the ones who mind the least and even thrive in your climate.

CULLING

The most common fate of culled birds is the table. Chickens can be harvested and processed on your farm for home use. State laws differ on how many birds you can sell to the public from uncertified processing facilities. Local farmers may have outlets to which live birds can be sold, usually for ethnic live bird markets. Ask around.

There are several methods for butchering your own poultry. Killing cones are one humane way to kill and bleed chickens. The bird is placed head down in the cone, which restrains it. The flow of blood to the head may have a sedating effect. The throat can be cut and the bird bleeds out naturally.

Chickens can also be killed quickly by hand. Holding the feet in one hand and the neck in the other, you twist the neck to dislocate or break it. Hold the bird until it is quiet, a minute or so. The blood will collect at the point of the break and can be drained out without making a mess. This method requires some experience to do efficiently. Work with someone who has done it successfully. A hand-held tool is also available but works best on small chickens.

The carcasses are dipped in vats of boiling water to scald them, making it easy to pluck the feathers. The carcasses are then gutted and cleaned. Cooling them in ice water speeds the freezing process.

Chickens can also be skinned, avoiding plucking entirely. Make a slit at the base of the neck where it joins the breast. Peel back the skin, cutting at the knees and removing the head. At the vent, make a circular cut and remove the innards.

Have a big enough freezer to safely keep as much poultry as you plan to acquire.

Culls don't necessarily have to be killed. Hens that lay well but are lacking in some other characteristic are welcome for their eggs. They may be perfect birds for fanciers who want fresh eggs but don't plan to breed or show their birds. Individuals of an unusual breed may be welcome at the local zoo or petting zoo.

Processing

The National Center for Appropriate Technology and the National Sustainable Agriculture Information Service have publications that can help guide you in making decisions on how and where to process your poultry.

Producing ready-to-cook poultry involves:

Preslaughter catching and transport

Immobilizing, killing, and bleeding

Feather removal by scalding and picking

Removal of head, oil glands, and feet

Evisceration

Chilling

Cutting up, deboning, and further processing

Aging

Packaging

Storage

Distribution

From *Small-Scale Poultry Processing* (2003) by Anne Fanatico

Cockerels can be caponized, or castrated, and raised as table fowl. They grow larger and faster and the meat is considered tender and tasty.

Because the sex glands are internal, a surgical incision and procedure are required to remove them. Caponizing kits are available online and at feed stores. Many small-flock owners learn to do it and make use of extra roosters this way. The procedure is performed on young cockerels, typically three to five weeks old, not older than eight weeks. The testes are small and firm at that age, making complete removal more likely to be successful. The testes in older birds get larger and softer, making accidental tears more likely. Partial testes left behind will continue to produce testosterone, so the bird will not be a full capon and may develop a rooster's comb and wattles. They are harvested at four and a half to six months old.

Capons are a specialty roasting fowl traditionally associated with holiday feasts.

Corallina Breuer

PROFESSIONAL PROCESSING

If you want to sell large quantities of poultry meat to the public, you must have your chickens processed at state or federally certified processing facilities. The plants are monitored for cleanliness and safety, both for workers and for consumers.

USDA-inspected processing facilities generally process up to 250,000 birds a day from large producers. The convenience of being able to process many birds of the same size, so that resetting the machinery and accommodating small batches does not interrupt production, keeps costs low. Large poultry operations are becoming more vertically integrated, with many of the producers owning their own processing plants.

Processing at state-licensed plants can cost between $1.00 and $1.75 per bird, adding significantly to the cost and limiting the market for small-flock birds. Processing remains a barrier to small-flock owners. However, with increasing interest in poultry and determination to support sustainable agriculture, solutions are emerging.

This killing cone restrains the bird and allows humane dispatch. Cones can be hung over a sink or receptacle to catch blood. The chicken is quickly killed and the meat is not bruised.

This Mobile Processing Unit has crates to hold the birds to be processed in the center. The killing cones are at the end. The scalder is on the right and the plucker on the left. At the far left are three workstations on a stainless-steel table.

Judith Kleinberg

The plucker is between the killing cones and the workstations. The unit is open to let the heat from the propane-heated scalder disperse. A tarp can be raised over it for shade.

Judith Kleinberg

One such solution is the Mobile Processing Unit (MPU). A pilot unit was built with grants from Heifer Project International and the USDA-Natural Resources Conservation Service Graze-NY program by the Resource Conservation and Development Council of South Central New York in 1995. Built on the frame of a reclaimed mobile home for about $3,000, an MPU can be towed by a half-ton pickup to small-flock locations. Producers pay a fee to use it. Kentucky has a state-approved MPU that rents for $50 a day after required training.

Custom processing plants may have the equipment and be able to process chickens for you. Contact your state department of agriculture to find out about small-flock poultry processing options in your state.

Jim McLaughlin, who was coordinator for the original New York project, now consults with pastured poultry producers. Having worked with state and federal regulators, he supports established producers and encourages novices to get started. The food-safety justification for state or federal certification has not, in his experience, proven necessary to assure clean, safe poultry.

"The people who raise the birds know the birds. They are selling to friends, family, and the Rotary Club. They know the importance of cleanliness," he says. "I have never heard of a single person who got sick from a bird raised on a small farm."

Legal Exemption

All states allow farmers to process small numbers of poultry for private use and small-scale trade. Facilities that accept such small numbers of birds for processing are referred to as exempt, because they are free from state and federal regulation.

Some states limit small-scale processing to one thousand birds a year. Some allow as many as twenty thousand. Consult your state law to determine what restrictions apply.

This stainless-steel double sink holds several birds for rinsing and initial chilling. Two people working the unit can process as many as twenty birds an hour.

Judith Kleinberg

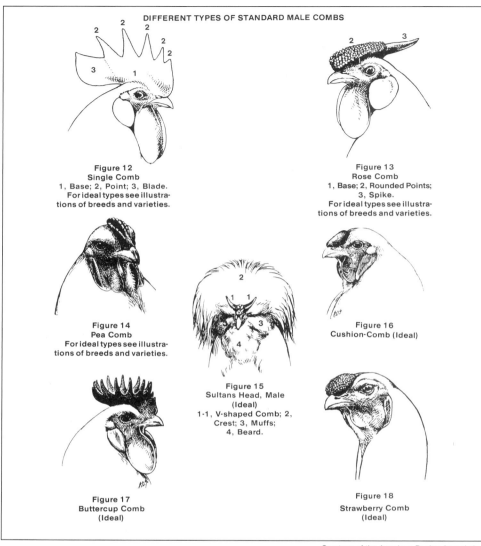

DIFFERENT TYPES OF STANDARD MALE COMBS

Figure 12
Single Comb
1, Base; 2, Point; 3, Blade.
For ideal types see illustrations of breeds and varieties.

Figure 13
Rose Comb
1, Base; 2, Rounded Points;
3, Spike.
For ideal types see illustrations of breeds and varieties.

Figure 14
Pea Comb
For ideal types see illustra-
tions of breeds and varieties.

Figure 15
Sultans Head, Male
(Ideal)
1-1, V-shaped Comb; 2,
Crest; 3, Muffs;
4, Beard.

Figure 16
Cushion-Comb (Ideal)

Figure 17
Buttercup Comb
(Ideal)

Figure 18
Strawberry Comb
(Ideal)

Courtesy of the American Poultry Association

Cock's combs can take many shapes and sizes. Comb styles are inherited, so breeders pay close attention to them in selecting breeding birds. Comb defects and disqualifications can be difficult to breed out.

Corallina Breuer

This Welsummer cock displays a comb that meets the standard for his breed. It is medium-sized, firm and upright, and divided into five distinct points, and the back does not touch the line of the skull and neck.

Brenda Ernst

In some breeds, a comb that falls over to one side is acceptable in hens, such as this Cuckoo Marans. Breeders take note of standards in other countries for breeds that are not recognized in the United States.

Corallina Breuer

This Cuckoo Marans cock shows the typical feather pattern of this French breed. Breeders aim to perfect the color and pattern of their chosen breeds by meticulous care in selecting breeding birds.

This bantam chick is a cross between a Silkie and an Ameraucana. Crossbreeding can add desired traits, but other traits often come along. Selection is a delicate art, balancing the desirable against the unwanted.

CHAPTER 11
• •

SHOWING

If you fall in love with your poultry, or become more engaged with breeding and husbandry, or find you enjoy the company of other poultry enthusiasts, you should show your own birds. Attending shows is fun. Participating is even better.

Raising poultry can be isolating—just you and the chickens. Shows are an occasion for poultry enthusiasts at all levels to get together on common ground. You may find answers to questions and solutions to problems or be the source of help to others. Breeders exchange stock at shows, invigorating their flocks with new bloodlines.

Shows give the public an opportunity to see and enjoy chickens, something that isn't in their daily lives anymore. It's a place for novices to start and experts to get new ideas.

Shows include lots of poultry-support information, from feed dealers and other vendors to photography services and marketing outlets. Whatever direction your flock is leading you, someone at the show will be able to offer you help and the advice born of experience.

Shows often include special youth events, such as silent auctions, raffles, and quiz bowls modeled on game shows to test poultry knowledge.

Appropriate biosecurity measures ensure that disease won't be brought in or taken home from a show. Participating birds are required to be tested for pullorum disease and other tests may be required, depending on local regulations.

ATTENDING

The county or state fair usually has a livestock section that includes poultry. Anyone may attend, although exhibitors may be restricted. Attending a show is a great way to find out what is going on with poultry enthusiasts in your local area. One of the advantages of shows is that they give the general public an opportunity to see poultry alive and in full feather.

Cards on the cages provide basic information about the bird: breed, sex, and age. Cockerels and pullets are, respectively, males and females in their first year. Cocks and hens are one year of age or older. Numbers are used to identify owners, to avoid any personal influence on the judge's opinion.

Shows publish an Exhibitors' List identifying all exhibits and their owners, along with contact information. If you are interested in a particular bird or breed, don't hesitate to introduce yourself and ask questions. Chicken breeders are proud of their birds and eager to talk about them with an interested audience. Take advantage of the expert knowledge available at shows.

Courtesy of the American Bantam Association

The American Bantam Association (ABA) is the presiding organization for exhibition of purebred bantams. The organization sponsors its own shows and meets at American Poultry Association (APA) shows. The ABA publishes its own *Standard* and licenses its own judges. The ABA and APA work together on many issues relating to exhibition and purebred poultry, including a youth program.

Other shows are regularly held throughout the year, more than one thousand annually. Check the *Poultry Press*. Contact the local high school agriculture teacher or county extension agent. They may be able to tell you when and where local shows are held.

PARTICIPATION

Shows are organized by clubs and other associations, so becoming a member is the first step. The American Poultry Association (APA) and the American Bantam Association

Christine Heinrichs

Ayrling "Butch" Gunderson, APA-certified judge, demonstrates how to hold a chicken for showing. Some chickens have a natural temperament for showing and accommodate the public with calm grace. Others need practice. Chickens should always be removed from the cage and replaced headfirst to maintain control and avoid wing damage.

(ABA) certify poultry judges, so they are involved in overseeing any show at which poultry are judged by certified judges.

APA certification assures that the judge has met its standards. Certified judges are required to qualify for points toward its Master, Grandmaster, and other awards.

Specialty breed clubs generally hold their own "meets" at poultry shows where they award their own separate prizes.

Ribbons, plaques, and prizes are awarded. Monetary prizes are nominal, but it's gratifying to get recognition among your peers for your achievements.

Young people in FFA and 4-H have additional events. The quiz bowl at a local show gives students an opportunity to learn to think on their feet. Elimination competitions lead to national events that offer scholarships and other substantial recognition.

Showmanship classes place students in a one-on-one examination with the judge. The judge interviews the young poultry owner on poultry knowledge and the ability to handle his or her bird. It gives young enthusiasts an arena in which to show off their best birds and shine.

Get all the relevant paperwork assembled and submitted before the deadline. Make sure your entries are recorded and you will be permitted to show. Some shows allow walk-ins, but many require preregistration.

PREPARATION

Getting ready to show chickens starts as early as breeding decisions and hatching. Your best birds are the ones you will want to show, so the ones that catch your eye right from the start will be the ones you eventually select for the show coop.

Refer to the APA *Standard of Perfection* and ABA *Standard* as your guides to what the judge will be using to evaluate your birds. They include a section on defects and disqualifications and how points are calculated, so you can avoid entering a bird that is automatically disqualified by a shortcoming that might be a minor defect in one breed but a disqualification in another. For instance, vulture hocks—long, stiff, straight feathers pointing downward from the back of the lower thighs—are a disqualification in most breeds but are a requirement for Sultans and d'Uccle and Booted bantams.

Christine Heinrichs

The judge examines the head for color, size, and shape. He is looking at the beak, face, eyes, comb, and wattles for any defects or disqualifications.

Getting chickens accustomed to being handled starts when they are young. Work with any birds you plan to show by picking them up, holding them, petting them, and touching their combs, wattles, and feet until they are able to remain calm. If started as chicks, nearly all birds will adjust to handling. Some love it and some tolerate it.

Successful show birds present themselves well to the judge. You can encourage them by letting them practice being in a show-size coop and rewarding them with a treat when you approach. A bit of corn, bread, or soft cat food will convince the bird to enjoy the experience and be happy to greet the judge.

Practice poking your show birds with a judging stick, as the judge will do to get them to present themselves and be evaluated. Your birds need to be accustomed to a judging stick so they perform as expected, rather than be intimidated or hostile.

You may want to protect your show birds by vaccinating them against diseases they may encounter in your area. That may include Marek's disease, Newcastle disease, laryngotracheitis, Avian Encephalitis, fowl pox, fowl cholera, and infectious bronchitis. Vaccination against avian influenza exists but is not presently allowed by the USDA. State law varies in its requirements. The show secretary will inform you about which vaccines are required.

Anytime birds from various locations are gathered together, the possibility of disease transmission exists. Quarantine your show birds for two to four weeks after they return from a show.

Birds should be identified, usually by numbered leg bands. Most common are the bands that snap in place when the bird is first blood tested. Some leg bands require pliers and others slide together. Leg bands are applied between eight and twelve weeks of age, when the hind toe is still flexible and the band can be slid over it onto the shank. Cockerels need to be checked as they grow, to make sure the band is above the developing spur. If it stays below the spur, it can constrict the shank and become uncomfortable.

Preening

Chickens, like all birds, have a gland at the base of the tail, the uropygial gland, which secretes semisolid fatty substances. Chickens use their beaks to spread these secretions over their feathers and keep them in good condition.

This rich, oily substance is a mixture of waxes, fatty acids, fat, and water. It prolongs the life of feathers and keeps them moist and flexible. It inhibits growth of fungus and bacteria and discourages lice.

Chickens spend time preening, dipping their beaks in the secretions, and smoothing it across their feathers. They are often businesslike about it but appear to enjoy it, too.

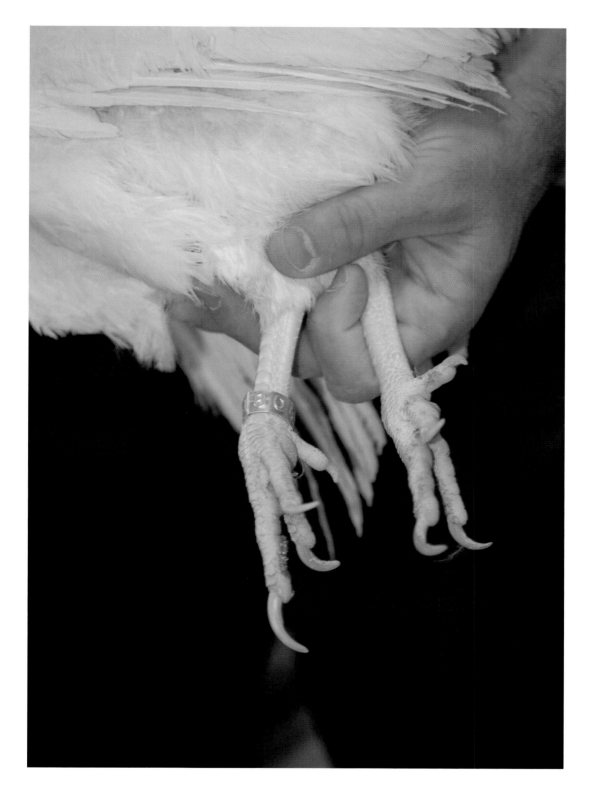

Metal leg bands permanently identify birds. While not required, they protect the owner from any confusion or misunderstanding about the ownership of the bird.

READY TO SHOW

Birds must be clean and neat. That may include shampooing, in the case of soft feathers and light colors. Use warm water and pet shampoo, keeping shampoo away from the head and eyes. Rinse well. Conditioner can be applied to smooth rough feathers and again rinsed well. Hard-feathered birds may only need their heads wiped off and feet scrubbed. Dried manure must be cleaned away from the vent area.

Scrub feet and legs with a brush and a powdered kitchen cleanser or hand cleanser. Most birds that are accustomed to being handled enjoy this. One chicken owner said her rooster "purred" when she scrubbed his feet. Follow with baby oil or petroleum jelly on the legs and feet. This treatment keeps them clean and improves appearance. Baby oil brings out the best in combs and wattles. Some exhibitors add a few drops of rubbing alcohol to the baby oil, but use it with care. Alcohol can burn and dry out these sensitive areas. Baby oil also discourages scaly leg mites, which would prevent you from bringing that bird to the show.

Bathing chickens should be done at least three days before the show, to allow time for the bird to get completely dry and preen her feathers back into shape. Smooth-feathered

Christine Heinrichs

The judge extends the wings to examine the 10 primary and 14 to 18 secondary feathers. The axial feather is between the primary and secondary flights. Covert feathers fill in the spaces.

The judge examines the feathers for color and condition. Feather-quality standards vary by breed. This Buff Orpington's feathers must be moderately broad and long and fit close to the body.

Christine Heinrichs

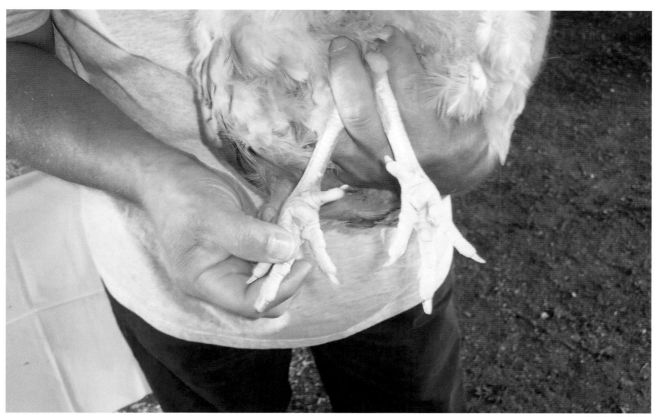

Christine Heinrichs

The feet and legs must be clean, so the judge can examine them for abnormalities, feathers, and color. Bowed legs, any deformity of the joints, and other abnormalities are disqualifications. Leg and foot color are important qualities.

Faking

Many techniques are allowed and even expected to prepare chickens for shows, but deceptive techniques, such as dying or plucking feathers or making any surgical changes to comb or wattles, is strictly unacceptable. Evidence of "faking" is grounds for disqualification.

Occasional small feathers growing from legs and feet of clean-legged breeds can be tweezed out. Presence of a feather follicle is grounds for disqualification, so tweeze only small, minor feathers. Too many feathers may be a consideration in deciding whether to cull this bird.

Plucking main feathers is grounds for disqualification. Some birds have one or two feathers the wrong color. If plucked, the new feather may grow in the correct color, but allow at least two months for this to happen.

Modern Game and Old English Game adult cocks are disqualified if they are not dubbed. Dubbing is the close trimming of the comb, wattles, and earlobes. Dubbing is a practice harking back to cockfighting, to remove soft tissue that an opponent could hang on to. It is usually done on chicks.

Poultry competition was cutthroat in the late nineteenth and early twentieth centuries. Faking feathers and other characteristics was a problem for exhibitions. In 1934, George Riley Scott wrote *The Art of Faking Exhibition Poultry*, which condemned the practice while giving precise instructions on how it could be done. Practices such as grafting perfectly marked feathers into the cut-off quills of undesirable ones was common. A facsimile edition of this historical curiosity is currently available.

breeds such as Rhode Islands, Brahmas, and Langshans may need more time to replace the oil that will be washed away along with the dirt.

Towel dry the bird in a smooth, firm movement from head to tail, pressing excess water out of the feathers. Don't scrub! Scrubbing can damage feathers. Place the bird in a warm, dry box or cage with clean bedding and allow to air dry. Wet chickens can get chilled and it can take 12 hours or more for them to get completely dry. Hair dryers can help, but too much heat can damage feathers.

TRAVELING TO THE SHOW

Have adequate cages or carriers for all the birds you will be taking to the show. Remember to plan for birds you intend to sell or exchange as well as the birds that will be shown.

Pet carriers are good. Cardboard boxes may suit your needs. Cut holes in them to ensure adequate ventilation. Depending on how long the journey is, you may have to provide food and water. Allow time to take travel breaks every four hours or so and offer water to your birds.

Waterers inside the cages may spill during the drive. Chickens can have food in their carriers, both scratch and greens. Cages with wire bottoms allow chickens to peck at grass without leaving their carriers but can risk toes during transit. Protect the cage bottom during the trip.

The trip should not be stressful or your birds will not show well. Keep them cool and calm. Some birds travel well in small groups, but some get fussy and fight. Make sure they will get along and you won't find broken feathers when you arrive at the destination.

SHOW ETIQUETTE

Be on time. With many exhibitors bringing multiple birds to a show, cooping in is hectic under the best conditions. Be courteous to your fellow exhibitors. Be patient.

Provide informational cards describing your birds. These can be helpful in explaining the history and background of your breed in general and your birds in particular. Shows are an excellent opportunity to raise awareness of poultry.

Horst Schmudde

Horst Schmudde

On Cubalayas, multiple spurs are preferred. The heavily feathered lobster tail is required; it is longer than the rest of the body, carried low. Cubalayas are considered excellent meat birds. They are also good layers and the hens make good mothers.

Dubbing is required in Old English and Modern Game roosters but disqualifies any other bird. This Leghorn was dubbed as a chick to identify him as a member of a particular research flock.

Shows generally offer space for cages to display stock for sale. These birds are not vetted for health or condition, but they are often excellent. The owner is at hand and you can see what other birds he or she is showing.

Owners may be willing to part with prizewinning birds. If you see birds you are eager to add to your stock, ask the owner whether he or she is willing to sell or has other birds available.

Birds purchased at a show, like all stock new to your flock, should be quarantined two to four weeks before introducing them to the flock to avoid transmitting any disease they may harbor.

This Leghorn rooster's outsize comb has one point too many for perfection, but its bright red color and deeply serrated points are in his favor. Comb development can be influenced by temperature and other environmental conditions as the chick is growing.

Corallina Breuer

Never touch another person's bird. If you see a situation that concerns you, such as a bird prostrate or losing feathers, find the owner and inform him or her. Touching birds has overtones of interfering with the competition by breaking feathers or other nefarious activity.

Feed and water according to show advice. Some breeds, such as Old English Games, do not show well with a full crop. You may want to temporarily remove water. Reassure concerned members of the public that the chickens are not being deprived and will soon have free access to water again.

Do not interrupt the judge or attempt to converse with him or her in any way. This smacks of attempts to influence the judging. Stay away from the judging area until judging is complete.

Participate in your events and support your fellow exhibitors. A responsive audience is rewarding to exhibitors who have taken the trouble to put their best birds forward. Classes such as showmanship benefit from an audience, because part of the challenge is to be able to think under pressure. Be respectful. Be appreciative.

Stay until all prizes have been awarded. Give everyone the courtesy of admiration during their time in the winners' circle.

Be a gracious winner. Kind words of fellowship and encouragement are always welcome.

COOPING OUT

Help with any cleanup. Settle any accounts. Confirm addresses for any follow-up mailings. Volunteer to help with the next show. After these duties are done, pack up and coop out promptly. Settle your birds for the trip home.

At home, quarantine the birds who attended the show for two to four weeks, to avoid spreading any disease they may have picked up. Write up your news for the local paper. Pictures, especially of young people and their chickens, are welcome. Share your news with the community. Make a nice display of your ribbons and enjoy your successes. Learn from your mistakes. Give yourself a good rest before you do it again.

Phil Bartz

This Black-Tailed White Japanese bantam is a study in contrasts, from her bright red face, comb, wattles, and earlobes to her silvery white body and wings and black tail feathers edged in white. Her wing primaries, unseen, are also black edged with white.

Courtesy of Shagbark Bantams

This Serama cock is one of the smallest bantams, with a preferred weight of under 350 grams and a top weight not more than 450 grams. The Serama has been bred in Malaysia as a living work of art.

Corallina Breuer

Chickens love to preen, taking oil from a gland at the base of the tail and spreading it over their feathers. Feathers benefit from care. Depend on your chickens to preen themselves after a shampoo.

HEALTH CARE

Corallina Breuer

This Buff Orpington hen isn't sick; she's molting. Like all birds, she replaces her feathers periodically. Most chickens molt annually.

Good general nutrition and husbandry practices will nearly always keep your chickens healthy. The USDA found only 13 percent of backyard flock owners reported any health problems during the years, most of them minor.

Just as in humans, a long list of varied diseases and disorders can plague chickens, but most of the time, illness is caused by a few familiar bugs. Small-flock owners manage the occasional sick chicken on their own. Any sick chicken that does not respond within a few days to your ministrations requires professional veterinary attention, as does the rapid spreading of illness in your flock.

Commercial chickens living in crowded conditions are subject to many more diseases than backyard flocks. Commercial breeders sometimes accuse backyard flocks of harboring disease organisms. The fact is we are all surrounded by a variety of potential disease-producing organisms. Good general health and clean living conditions enable humans and chickens to resist infection and disease or, if infected, recover from it.

Backyard chickens are among the most biologically isolated critters in the world, in the sense of contacting microorganisms that can make them sick. However, there are some nasty bugs out there, as well as some that are annoying and debilitating. Recognizing symptoms and treating affected individuals early can save you money and heartbreak.

Always be alert to changes in behavior that may signal disease or infestation. Loss of appetite is a common symptom but difficult to notice in a flock. Stay on the lookout for other signs of illness, including swollen joints, diarrhea, swollen sinuses, difficult breathing, runny nose, scabby sores, and neurologic signs such as drooping wings or loss of coordination. Even if prompt treatment can't save that individual, it can help you avoid losing the whole flock.

As prey animals, chickens notoriously avoid showing weakness. The first sign you see may be a dead chicken. But most things that afflict chickens are not that extreme and, with supportive care, they will recover completely.

GOOD HEALTH AND HUSBANDRY

A good diet, clean living conditions, and lots of fresh air and sunshine are the best basics for a healthy flock. Poultry suppliers offer good vitamin and mineral supplements, such as Calf Manna pellets, Purina's Animax, and Nutrena Milk Plus.

Nutritional supplements, such as the beneficial bacteria found in active yogurt cultures and *Lactobacillus acidophilus* from the local health-food store, are good for chickens. Beneficial bacteria are called probiotics. They can be added to the diet as a regular supplement or reserved to replenish the digestive tract after antibiotic treatment, which wipes out both pathogenic and beneficial flora.

If one of your chickens starts acting sick and you suspect an infectious disease, quarantine the bird from the rest of the flock. If other birds have been exposed, confine them together. Chickens are flock birds that are stressed by isolation. If you don't suspect something infectious, put a couple of her friends in with her.

Sick birds benefit from higher protein feed, as much as 30 to 40 percent protein. As a regular feed, that much protein would cause kidney problems, but in sick chickens it helps build cells and hasten recovery. Game bird or pheasant starter is good. Chickens love cat or kitten food, which is high in protein.

You may be able to tempt birds who are off their feed due to illness with treats like creamed corn. If she won't eat at all, you may have to syringe-feed and water her. Squirt it into her mouth gradually, giving her time to swallow. Put a small amount of feed mixed to a loose paste with water or milk in her mouth. She will probably start eating for you. She needs to take some nourishment. Sick chickens can die of starvation.

If you have quarantined her in a cage, give her time out in a yard a couple of times a day. She needs fresh air and exercise, even if she's not feeling well. Encourage her to walk around and flap her wings. Try leading her around with bits of a food she loves. Encourage her to fly up onto your lap or shoulder.

Do not hold her upside down in an effort to get her moving. Even if she is a pet, don't handle her too much. It can be stressful even for those that are used to it.

FINDING A VET

It's worth your time to locate a veterinarian before you need one. Most vets aren't familiar with the disorders that afflict chickens and can't get up to speed quickly. Bird vets usually prefer to focus on exotics or cage birds, which are different from chickens, whereas livestock vets usually work with large mammals and avoid poultry.

Contact the Association of Avian Veterinarians for a local referral and inquire as to whether that person handles chickens or if they can refer you to someone who will. An experienced poultry keeper may be able to offer a recommendation. You may have to persevere a bit, but identifying someone before you need them is worthwhile. You may want to vaccinate your birds or have other questions on which you'd like a medical perspective.

Corallina Breuer

This Barred Rock rooster has all the indications of perfect health in his smooth black and white feathers, upright sprightly stance, and clear eyes. Chickens are generally healthy and recover from minor infections quickly.

INJURIES

Perhaps the most likely need you will have for a vet will be the result of a predator attack. Owls, opossum, raccoons, foxes, and other predators are always on the prowl for tasty chicken and eggs. Despite your careful plans, predators who study your yard may find a way in, and you may find yourself with injured fowl.

You can handle most minor injuries and infections yourself. Keep a triple antibiotic on hand, along with bandages and tape. Vet Wrap is difficult even for a determined bird to remove.

Nail clippers, scissors, and a nail file are useful for trimming spurs and nails to avoid injuries. Beaks can also be trimmed with the clippers. Crossed beaks are a genetic defect that should not be allowed into any breeding program, but with supportive care, individuals can be successful egg, meat, or companion birds. The beak can be carefully trimmed or filed to a better fit. Keep feed deep so that she can get enough to eat.

ADMINISTERING MEDICATION

Medication can be added to food or water, injected, or applied by controlled drench. Controlled drench is a measured dose of liquid medications applied by syringing it down the chicken's throat.

Putting medication in the food or water is the simplest but makes it impossible to know for certain what dose the chicken actually received. During hot weather, the bird might get too much medication. Unless the sick bird is isolated, other birds would also get the medication. In the case of antibiotics, insufficient doses given indiscriminately to the entire flock can create resistant strains of bacteria that could become chronic problems.

Your vet may be willing to teach you to inject your own chickens, in the event they need injected medication such as antibiotics. Intramuscular injections (IM) are given in the breast muscle. Hold your chicken securely in one hand, resting on your forearm. Roll her over to present one side of her breast. Swab the area with alcohol on a cotton ball and inject firmly past the skin into the muscle. Pull back on the syringe to make sure the needle isn't in a blood vessel, not likely in this area. If clear fluid appears in the syringe, inject the medication or vaccine. With some practice holding your chickens and injecting into an orange, you can manage this.

You may want to keep some veterinary antibiotics such as Amprol or Corid, administered in feed, on hand for treating infections such as coccidiosis should they arise.

The National Poultry Improvement Plan (NPIP) monitors egg-transmitted diseases such as pullorum disease, fowl typhoid, mycoplasmas, and *Salmonella enteritidis*. They are prevented by culling infected birds and not allowing them to pass on the infection in

Cell Migration

For large, serious injuries that leave one or more of your birds with gaping wounds, cell migration may be able to save them. It's an intensive method requiring lots of care, but it can restore skin and feathers. And it's the only way to replace skin that includes feather follicles, preventing bare patches that may become targets for pecking.

Predator attacks can result in severe wounds that leave too little skin to be pulled together for stitching. The principle of cell migration is to allow the remaining skin to regenerate across the wound, growing new skin rather than scarring.

The bird needs to otherwise be in good health and worthy of your time and attention. Keep the wound scrupulously clean and moist, covered in triple antibiotic ointment and a protective bandage. The bird needs to be kept away from other birds for the duration. Change the bandage daily to inspect the wound and remove all scabbing and dried material. An oral antibiotic followed by supportive probiotics will help the bird recover. Gradually, new skin will grow from the edges of the wound to close it.

their eggs. Your vet or extension leader can refer you to individuals who are certified to test for these diseases.

Nearly all pathogens—the microorganisms that cause disease—and pests are specific to one or a narrow range of species. So your chickens won't catch your cold or get fleas from your dog.

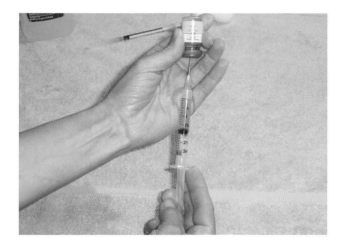

Many small-flock owners manage routine vaccinations against diseases such as Marek's themselves. The vaccine is freeze-dried, so it first must be rehydrated. K. J. Theodore provides a step-by-step guide on her Shagbark bantam website.

Hold the chick facedown. Use your thumb and forefinger to separate the down on the back of the neck. Wipe it with an alcohol swab. You will be able to see tiny veins beneath the skin.

Insert the needle toward yourself, just under the skin, not into the muscle. Lift the tip of the needle slightly to be certain it is correctly placed.

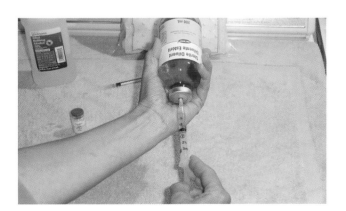

Draw up the rehydrated vaccine. Each chick will receive 0.2 cubic centimeters (cc) of vaccine, so the 1-cc syringe can hold enough to vaccinate five chicks. The same needle can be used for all the chicks you are vaccinating that day. Wipe it with an alcohol swab between injections.

Inject the 0.2 cc of vaccine. It will form a bubble under the skin. The chicks will be a bit disoriented at first. Watch them carefully for 24 hours to make sure they stay under the brooder and on their feet.

All photos courtesy of Shagbark Bantams

NATIONAL POULTRY IMPROVEMENT PLAN

The NPIP is a USDA program directed at the control and elimination of poultry diseases that can be transmitted through the reproductive process.

Salmonella pullorum and *S. gallinarum* (also called pullorum disease and fowl typhoid, respectively) are bacterial diseases that can devastate chickens. Young birds are most at risk, but birds of any age can be affected. Survivors of these infections remain carriers for life and can become sources of infection for other birds. As breeders, their hatching eggs are contaminated. Chicks are hatched with the infection and frequently die.

The infection may spread when chickens breathe contaminated dust or come into contact with down from infected poultry or with other material in the incubator, shipping box, brooder, or pen that has been touched by an infected bird.

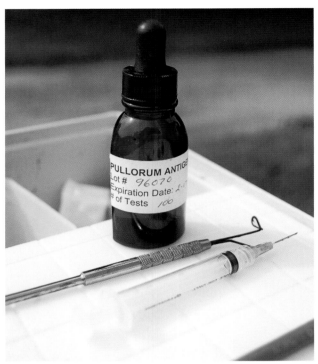

Corallina Breuer

State departments of agriculture offer training to become certified as a pullorum tester with the National Poultry Improvement Plan. The equipment and supplies required are available at poultry supply houses.

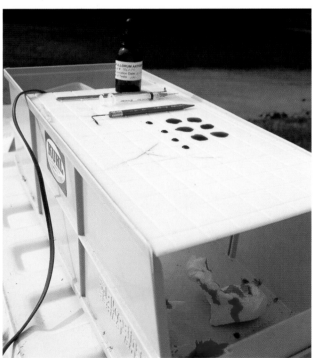

Corallina Breuer

All pullorum testing materials must be treated as possibly infectious. The diseases rarely appear, but an outbreak would be serious for both small flocks and commercial operations. After testing, flocks are certified clean for a year.

Corallina Breuer

This Barred Rock is having its blood sample taken. Using a needle, the test pricks a vein to produce a small pool of blood.

Corallina Breuer

The blood is collected on a loop for transfer to the testing antigen. Only a small drop of blood is needed to test for the antibodies. Not every bird in a flock is necessarily tested.

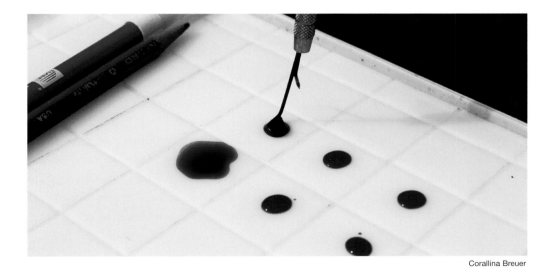

Corallina Breuer

The blood is mixed with the antigen to determine the presence of antibodies. If the antigen clumps together, antibodies are present. The presence of antibodies indicates that the bird has been exposed to the bacteria.

Blood testing adult breeding birds, biosecurity, and sanitation have all but eliminated the disease in the United States.

Testing for these diseases is required for exhibition and for sales. Training to become an NPIP-certified blood tester is available. Many small-flock owners are certified. If you cannot find someone, blood samples can be sent to NPIP-approved laboratories.

BACTERIAL AND FUNGAL DISEASES

STAPH INFECTION

The common germ *Staphylococcus aureus* is normally present in chickens but can infect broken skin from scratches or excessive moisture, causing a staph infection. The chicken will have swollen joints or footpads or a sore oozing yellow pus. Chicks may die with swollen abdomens and scabbed navels.

Antibiotics such as erythromycin, available from your vet or the feed store, will clear up a staph infection. Avoid it by eliminating sharp objects from the chicken house, removing wet litter, and fixing dripping waterers.

COLIBACILLOSIS

The *Escherichia coli* bacteria that causes this infection—also called coligranuloma, colisepticemia, or mud fever—is a normal resident in the digestive tract of chickens and commonly found in manure. Birds that get sick from *E. coli* lose weight and stand around with ruffled feathers. They may have enlarged and swollen navels, and feathers will be pasted around their vents from the diarrhea.

Birds don't transmit colibacillosis to each other. They catch it from dirty litter. It can be treated with mycin antibiotics.

Keep the chicken house clean and dry. After an episode of colibacillosis, disinfecting the chicken house with bleach or other disinfectant is a good idea.

AVIAN MYCOBACTERIOSIS

This chronic disease, also called avian tuberculosis, spreads slowly but can gradually kill your birds. It causes nodules called granulomas to form in the intestines, spleen, liver, and bone marrow.

The bird gradually gets skinnier and weaker. It occasionally has diarrhea. It may drop dead without warning.

Birds catch avian mycobacteriosis by ingesting contaminated feed, water, or litter. If one of your birds turns up with it, you will have to test the rest of your flock to find whether any others are infected, then separate them from the rest of the flock for treatment. Some decide to sacrifice the infected birds instead of trying to save them; commercial birds are not treated.

The chicken house must be thoroughly cleaned and dried. The bacteria, *Mycobacterium avium*, can resist disinfectants. Use household bleach, Oxine, or other poultry disinfectants to make sure it is gone before using the structure again for chickens.

AVIAN MYCOPLASMOSIS

This chronic respiratory infection, commonly called chronic respiratory disease, is a major problem for the commercial poultry industry. It is monitored by the NPIP. Chickens may lose their appetite and reduce egg-laying. They may cough and sneeze. Their eyes and noses will run.

Avian Mycoplasmosis is caused by *Mycoplasma galliseptum* bacteria. Chickens catch it from each other the way we catch colds, from breathing in the germs. They also transmit it through their eggs. This is not a problem for humans who eat them, but it's impossible to eliminate in the chickens. There is no treatment that cures Avian Mycoplasmosis.

An effective vaccine exists, but it is expensive and not packaged for small flocks. Carriers can be identified by testing and eliminated from the flock.

ASPERGILLOSIS

Aspergillosis, also called brooder pneumonia, is not common in small flocks that have access to fresh air and sunshine. Birds catch this fungal infection from breathing *Aspergillus* spores generated in infected litter. It's a common organism, present in every chicken coop. Birds don't catch it from each other, but the organism can penetrate eggshells and broken, infected eggs will further contaminate the environment. Keeping the chicken house and yard clean is the best preventive.

Birds afflicted with aspergillosis gasp and struggle to breathe. They often breathe rapidly. Once they are showing these signs, they are likely to die. They may show neurologic signs as well, such as uncoordinated movements or partial paralysis, if the infection has gone to the brain. If the eyes are infected, they will be opaque.

The chicken house will need to be cleaned and disinfected and the air ducts cleaned out if you have them. Make sure litter is clean and dry. Clean and fumigate eggs before setting if you have a history of problems with aspergillosis.

INFECTIOUS LARYNGOTRACHEITIS

This respiratory disease is highly contagious, especially in confined flocks. Infectious laryngotracheitis (LT) is caused by a herpes virus and spreads from chicken to chicken as well as via contaminated equipment.

Birds cough and gasp and have runny noses and moist sounds in their breathing. Their eyes are watery and congested. Sinuses around the eyes may be swollen. The sickest birds will have trouble breathing. Blood-stained mucus will show up on their wing and breast feathers from slinging their heads around to clear their throats. Death can result from suffocation due to blood and mucus in the throat.

Effective vaccines exist for LT. Vaccines take seven to ten days to be effective but can save your flock if you identify LT before too many birds are sick. Follow the vaccination instructions closely, as it is a live vaccine and high doses can cause disease symptoms. Unvaccinated birds should not be exposed to vaccinated birds for thirty days after vaccination, as the vaccinated birds may shed live virus.

No treatment cures LT, but broad-spectrum antibiotics can reduce secondary infections such as colibacillosis. LT can be mild, and birds can recover in two weeks. Recovered birds acquire natural immunity.

NEWCASTLE DISEASE

Chickens share their susceptibility to this group of diseases with all other birds. Newcastle disease (ND) is classified by its pathogenicity into three categories: velogenic, mesogenic, and lentogenic.

Antibiotic Feed

Antibiotic feed is a controversial subject. Antibiotics are one of the advantages of modern medicine, but adding them to feed has been linked to producing antibiotic-resistant strains of bacteria in commercial animals.

Antibiotics have a place in fighting infections that can kill or debilitate chickens, not to mention humans. Studies found that removing antibiotics from feed for organic chickens resulted in triple the infection rate of bacteria that affect humans. More chickens got sick, requiring higher levels of antibiotics for treatment.

Starting chicks with feed medicated with Amprolium or Bacitracin can give them an advantage in growing. The amount is small, which may work to the advantage of developing natural immunity by allowing the birds to develop mild coccidiosis and enteritis infections from which they quickly recover.

Velogenic strains are severe and cause high mortality. It spreads through the flock rapidly. Birds may drop dead without showing any symptoms. Birds may be depressed, experience increased respiration, and gradually weaken to the point of prostration. They may have a watery, greenish diarrhea. They may cough and gasp and have runny noses and eyes. The comb and wattles may turn bluish. The head may be swollen. If they survive the initial stage, they develop neurologic signs, such as lack of coordination or partial paralysis. Egg production drops and some eggs are deformed. These strains can kill more than 90 percent of the flock.

Mesogenic and lentogenic strains are common in the United States. The symptoms they cause are similar to the velogenic forms but less severe. Mesogenic ND usually causes only respiratory symptoms, although it may cause some neurologic symptoms as well. Mortality ranges from 5 to 50 percent.

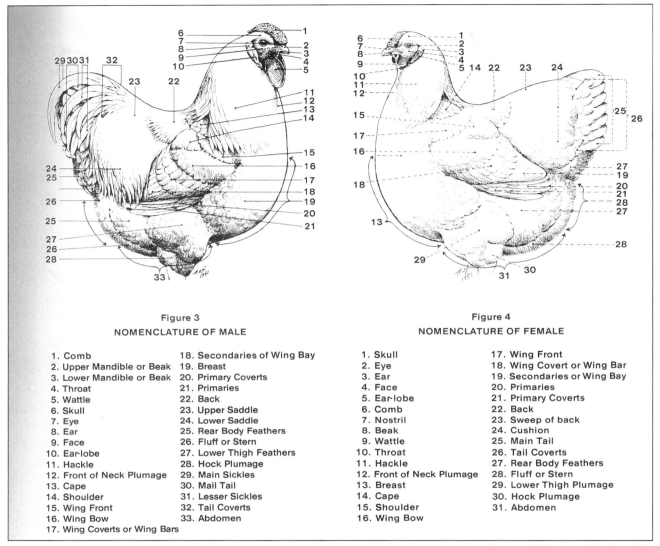

Figure 3
NOMENCLATURE OF MALE

1. Comb
2. Upper Mandible or Beak
3. Lower Mandible or Beak
4. Throat
5. Wattle
6. Skull
7. Eye
8. Ear
9. Face
10. Ear-lobe
11. Hackle
12. Front of Neck Plumage
13. Cape
14. Shoulder
15. Wing Front
16. Wing Bow
17. Wing Coverts or Wing Bars
18. Secondaries of Wing Bay
19. Breast
20. Primary Coverts
21. Primaries
22. Back
23. Upper Saddle
24. Lower Saddle
25. Rear Body Feathers
26. Fluff or Stern
27. Lower Thigh Feathers
28. Hock Plumage
29. Main Sickles
30. Mail Tail
31. Lesser Sickles
32. Tail Coverts
33. Abdomen

Figure 4
NOMENCLATURE OF FEMALE

1. Skull
2. Eye
3. Ear
4. Face
5. Ear-lobe
6. Comb
7. Nostril
8. Beak
9. Wattle
10. Throat
11. Hackle
12. Front of Neck Plumage
13. Breast
14. Cape
15. Shoulder
16. Wing Bow
17. Wing Front
18. Wing Covert or Wing Bar
19. Secondaries or Wing Bay
20. Primaries
21. Primary Coverts
22. Back
23. Sweep of back
24. Cushion
25. Main Tail
26. Tail Coverts
27. Rear Body Feathers
28. Fluff or Stern
29. Lower Thigh Plumage
30. Hock Plumage
31. Abdomen

Courtesy of the American Poultry Association

This diagram of the body parts of a White Wyandotte cock and hen was done by Arthur Schilling in the early 1900s for the APA. These drawings are so beautiful and precise that they are still used today.

Lentogenic strains are like a mild cold. Egg production drops off but will return to normal in a few weeks. Mature birds recover completely, but young birds may develop severe respiratory symptoms.

Live vaccines are administered through drinking water or by aerosol to the eyes or nose. Inactivated vaccines are injected and may be used after live vaccines to boost immunity. Regular boosters are needed to maintain resistance. A vaccine approved in 2003 is administered by machine to eggs before hatching. Broad-spectrum antibiotics can reduce secondary infections.

There is no treatment for ND, but supportive care to reduce stress on the birds can help them recover. If the outbreak occurs during cold weather, adding heat to the coop can reduce stress. Offering protein and vitamin supplements may also help reduce stress.

Clean and disinfect the chicken house. Maintain clean practices and an effective biological barrier if ND is in your area. The virus can be carried on clothing, shoes, vehicle tires, and poultry equipment. Keep dedicated clothes and boots to use only in the poultry house. Keep visitors to a minimum. Provide visitors with disposable booties. Add new birds only from uninfected flocks after keeping them quarantined for thirty days.

MAREK'S DISEASE

This herpes virus disease causes tumors. Tumors can grow in any organ; when they grow in the nerves, paralysis results.

Birds are infected by breathing the virus in the air. It circulates in the air from infected feathers and litter. Infected birds shed the virus through feather follicles. Recovered birds can remain infectious as long as eighteen months. Insects such as darkling beetles may also carry the virus from place to place on their feet.

Chicks can be vaccinated at any age, but the longer you postpone vaccination, the more time they have to contract the disease. Chickens can also be vaccinated as adults, and some veterinarians recommend giving the adults in your flock a booster every year when you vaccinate the chicks.

There is no treatment. This disease is a good reason to know who you get stock from. Stock from commercial hatcheries is safe because they are meticulous in guarding against Marek's disease. Vaccination for chicks is generally available for a penny or two each. If you hatch your own, you can vaccinate them yourself at one day old.

FOWL POX

Fowl pox occurs in two forms: cutaneous and diphtheritic. The virus can infect birds through any open sore. Mosquitoes can transmit the disease after biting an infected bird, and other insects can transmit it to birds through their eyes. Birds may be infected through the mucous membranes of the mouth and upper respiratory tract even if there is no injury to the membranes.

The cutaneous form, called dry pox, causes nodules that turn scabby on the comb, wattle, eyelids, and other unfeathered areas. The scabs eventually fall off.

The diphtheritic form, called wet pox, causes sores on the mucous membranes of the mouth, esophagus, and the upper part of the trachea. The sores may grow to join in a diphtheritic membrane covering the sores. The sores may make it difficult for the bird to eat, drink, or even breathe. As a result, birds can suffocate or starve to death.

Any scabby sores should be suspected of being fowl pox. There is no specific treatment. If your flock is infected, support them with good management to reduce stress until they recover. Keep them warm but ventilate the coop well with fresh air. Keep clean, fresh water available. Offer high-protein feed and vitamin and mineral supplements. Keep the litter clean. Antiseptic ointments may speed healing.

Effective vaccines exist for fowl pox. Tissue culture vaccine can be used on chicks as early as one day old. Chick embryo vaccine can be used on birds four weeks old and older. As with infectious laryngotracheitis vaccine, fowl pox vaccines are live virus vaccines and vaccinated birds may shed live virus for as long as thirty days.

INFECTIOUS CORYZA

Coryza is like a chicken's cold, otherwise known as roup, giving the bird difficulty breathing; noisy, crackling breathing called rales; and a thick, unpleasant nasal discharge. The face may be swollen and the eyelids may be stuck shut. Chickens may have diarrhea.

They catch it from other sick birds, but the severity of symptoms varies a lot. Your chicken could catch it from a new chicken being added to the flock or at a poultry show.

Water-soluble antibiotics such as sulfamethoxine are the best treatment. Vaccines exist for chickens on farms that have a history of coryza.

ASCITES

Ascites is congestive heart failure in chickens. It shows up at three to five weeks of age in Cornish/Rock cross broilers who had previously been fine. Some birds may drop dead. The afflicted birds are smaller than expected, reluctant to move, and have trouble breathing. Their abdomens are distended. Combs and wattles are pale or bluish in color.

Ascites occurs in breeds and hybrids that exhibit rapid growth and muscle development, resulting in fast conversion of feed to meat. Their meaty frames have more muscle-to-lung capacity than traditional breeds.

Ascites can be avoided by limiting feed, either through offering a limited amount in feeders or covering the feeders

In good health, this Ameraucana hen is ready to step out into the world and take her place in the pecking order. Sick chickens can be the object of pecking by other members of the flock.

Corallina Breuer

for one to three hours a day. Eliminate light a few hours each night. Maintain good air quality and ventilation, especially during the winter.

If the flock is losing more than 2 percent of its birds from ascites, investigate complicating factors such as high sodium levels in the feed or water, vitamin E or selenium deficiency, respiratory infection, or coal tar toxicity.

PARASITIC DISEASES

COCCIDIOSIS

Coccidiosis is the most common poultry disease problem, but it is nearly unknown in chicks raised by hens. Coccidia are a type of protozoa. Chickens are susceptible to infection by at least eleven species of the genus *Eimeria* coccidian. *Eimeria* is a genus of Apicomplexan parasites that includes various species responsible for coccidiosis. Infected chickens bleed intestinally, resulting in pale combs and wattles. Chickens typically look sick and depressed, while shivering with ruffled feathers. There may be blood in the chickens' droppings. Laying drops off.

Coccidiosis is caused by up to nine species of Coccidia protozoans. Those that infect chickens do not infect turkeys, and vice versa. At least two species are usually involved in infections. Because of varying vulnerability at various points in their life cycle, Coccidia can be attacked in several ways.

Live oocyst vaccines are available for six species of *Eimeria*. Vaccines are administered through food or water, exposing the chicken to a mild infection in order to develop immunity. Medicated chick starter allows mild infections that result in immune adults.

Anticoccidial drugs kill or decrease growth of the disease at one to four days of infection. Because Coccidia are likely to develop resistance to drugs, products should be rotated. Use one product, such as Nicarb or Coban, in the starter ration and change to a different drug, such as Avatec, in the fifth week.

Amprol can effectively control Coccidia, if given for two or three days and followed with another dose in two weeks. Poultry owners should change the pen or remove and destroy all litter.

ROUNDWORMS

Young birds under three months of age and lightweight breeds, such as Leghorns, are more susceptible to round-worm infection, or ascariasis, than heavy meat breeds. Infected birds are depressed, lose weight, grow slowly, and have diarrhea. Egg production goes down.

Deep litter reduces exposure to the worm eggs, which the chickens have to ingest to be infected. Clean up between flocks. Treatment with hygromycin B or coumaphos is effective but is not approved for layers.

Piperazine is a wormer given in drinking water. Some veterinarians recommend giving it preventively, but evidence is emerging that worms are becoming resistant to piperazine. Follow manufacturer's directions. Pasture rotation will help control all kinds of worms. Grazing different species on pasture, or letting pastures lie fallow until the eggs die, will reduce infestations.

CECAL WORM

The protozoan *Heterakis gallinarum* that causes blackhead in turkeys can be harbored in the eggs of this worm, becoming a possible source of infection in turkeys raised in close proximity to chickens that become infected with cecal worm. Cecal worm can also be transmitted by earthworms.

As with roundworms, deep litter reduces exposure to the eggs of cecal worms. Hygromycin B or coumaphos is also effective against cecal worm, but it is not approved for layers.

EXTERNAL PARASITES

Mites are arachnids that have eight legs, compared with the six legs of lice. The northern fowl mite spends most of its life cycle on the bird, but it can live off of the bird in crevices in the hen house for two to three weeks. Mites are

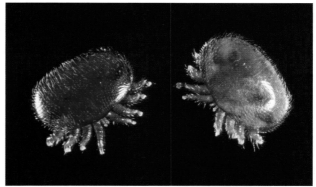

Courtesy of the USDA ARS Image Gallery

Mites are relatively slow-moving arachnids. Their eggs are found along the feather shaft.

Diatomaceous Earth

Many flock-owners swear by diatomaceous earth (DE) for keeping their birds clean of lice, mites, and internal parasites. This soft, white chalklike rock is pulverized for consumer use and can be applied directly to the chicken or sprinkled throughout the chicken yard to aid in dust-bathing.

DE is composed of fossilized diatoms, a hard-shelled form of algae. It is highly porous and absorbs more than its weight in water and oils. Food-grade DE is used to keep foodstuffs such as flour and pancake mix dry and clump-free. Medical-grade DE is used to worm humans and animals. The substance is put to use in many ways, including cat litter, insecticide, toothpaste, metal polish, swimming pool filters, and cleaning up toxic spills.

The most abrasive form of DE contains a high percentage of crystalline silica. Its razor-sharp microscopic edges make it hazardous to breathe. Dust masks are recommended for those who come in contact with it, especially the forms used in swimming pool filters.

Although the kind recommended for chickens has a low percentage of crystalline silica, some, such as Kim J. Theodore of Shagbark Bantams, who writes on poultry health, doubt that it can be entirely innocuous. Whether it can remain in the lungs or damage the intestines has not been proven but is a risk. Check it out and make your own decision.

Courtesy of My Pet Chicken, LLC

This chicken is giving herself a dust bath, working the gritty sand into her feathers and down to her skin to remove parasites and keep her feathers in good condition. She will conclude with a good shake and preening.

Tobacco and Pennyroyal

Tobacco has a long history as a medicinal for many ills of humans and animals. Nicotine is a powerful poison, making tobacco an effective parasite control. Early American colonists noticed its curative qualities, boiling it to make a potion to apply to "cutaneous distemper, especially the itch," as noted in the *Encyclopedia Britannica's* first edition in 1771. It was also used to cure the mange in dogs.

Adding tobacco leaf stems to henhouse nesting material will eliminate mites and lice. A nice handful of chewing tobacco serves the same purpose. It won't hurt your birds if they eat a bit and may help control worms.

Pennyroyal is the smallest of the mint group and has a pleasant and noticeable aroma. It gets its scientific name, *Mentha pulegium*, from *pulex*, the Latin word for flea. The Roman naturalist Pliny the Elder records that the Romans used it against fleas.

Pennyroyal makes another excellent addition to nesting material that will keep your chickens free of lice and mites.

Alvin Simmons, USDA

Tobacco grows easily in most climates. This wild tobacco and domestic tobacco are both effective for discouraging infestations.

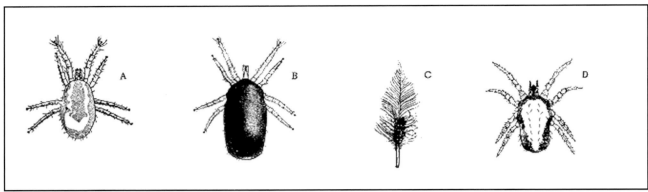

The northern fowl mite can be a destructive pest, but it is controllable. Pyrethrin is a natural botanical that kills it.

more common in cool climates. The chicken mite is a warm-weather pest, coming out of cracks in the henhouse to bite and suck blood at night.

Mites are easiest to see in the chicken's vent area, which attracts them because of its moistness. They may also appear on other parts of the body, such as the ears, the thigh, the neck, and the top of the head. Tiny black scabs may alert you to their presence. Light-colored feathers may show darkening due to mite feces.

Mites cause a general decline in flock health and vitality. They can deplete blood to the point of anemia. Check your birds once a month, more often in winter, for mites. A serious infestation can kill birds in a matter of days. If you find mites on one bird, check the rest of the flock. They may all have them and require treatment.

Minor infestations will clear up with two dustings of Carbaryl, or Sevin, powder and poultry/livestock dust. Two treatments, three days apart, are needed to break the reproductive cycle of mites. Poultry owners should remove the old bedding, dust the floor with the Sevin and poultry/livestock powder, and replace it with fresh bedding. Some chewing tobacco added to the bedding will control mites and lice.

More serious infestations require medicated shampoos and possibly injected medications. Consult with your veterinarian. The first step is to shampoo with a dog flea-and-tick shampoo, massaging down to the skin, and then rinsing out completely. Use care to dry the feathers as thoroughly as possible with a hair dryer. Your chicken may do better drying overnight in a carrier in the house unless it is early on a very warm, dry day. Don't let chickens get chilled. Treat any sores with triple antibiotic ointment. Dust chickens again in three days. The vet may also recommend another injection of medication at that time.

SCALY LEG MITES

Scaly leg mites live under the scales on the legs and feet of chickens and may infest the comb, wattles, and beak. The scales on the legs look like they are pointing outward instead of lying flat because of the mite burrowing into the keratin, causing a thickening of the scales.

Applying petroleum jelly to the legs, feet, and other affected areas is thought to suffocate the mites. Be persistent, and repeat the treatment in two weeks. Be patient, as healing the legs takes time. Severe infestations can cause permanent scarring of the legs.

Permethrin spray will kill the mites. It or other preparations such as VetRx can be used directly on the legs before applying petroleum jelly. It should be used on the hen house, coop, and any other facilities your chickens use to kill all mites and eggs.

LICE

Lice are tiny, wingless, flat-bodied insects that spend their whole life cycle on the bird. They feed on dry skin scales, feathers, and scabs. They do not suck blood directly but will eat blood oozing from irritated skin. Several different kinds infest chickens and more than one may infest a chicken at the same time. As with most other parasites, the lice that infest chickens will not infest humans.

Chickens pick them up from other chickens—another good reason to quarantine any new birds for two to four weeks before adding them to the flock. Then you can treat new chickens and their living quarters without involving the entire flock.

An infested bird's feathers may be dull from lice damage. Her feathers may look rough and moth-eaten. You may be able to see lice crawling around her vent area or see the eggs, called nits, attached to feathers.

Sevin powder kills the lice. The henhouse must be treated along with the birds. In severe cases, a pyrethrin-based spray, in the kitten dosage, can be used to reduce the number of lice, and then finish off with Sevin. Treat chickens by placing their bodies in a garbage bag with the chemical powder, holding it closed around the neck with the head outside the bag, and shaking the powder through the feathers. Wear a mask to avoid breathing the powder. The lice life cycle is about two weeks, so repeat the treatment in two weeks.

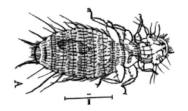

Courtesy of the USDA

Keep an eye out for infestations of poultry lice. A severe infestation can even kill chickens.

OTHER THREATS

Ironically, the greatest threat to birds may be from government agencies that require eradication of healthy birds to eliminate possible disease reservoirs. The USDA's National Animal Identification System (NAIS) aims to trace future infections. Livestock owners are banding together to question the value and efficacy of these programs. Farm for Life, a nonprofit that advocates for sustainable lifestyles based on local communities, is organizing opposition to the mandatory property and animal surveillance program as overly intrusive and unconstitutional.

Participate by joining other livestock owners in organizations, such as the SPPA and specialty breed clubs, to lobby for common interests.

AVIAN INFLUENZA

By now, everyone knows that Highly Pathogenic H5N1 Avian Influenza, otherwise known as bird flu, is killing birds and occasionally infecting and killing humans. Inevitably, human cases are connected to the possibility of a human influenza pandemic, although the scientific indications suggest that this possibility is remote. You and your neighbors want assurance that your birds are safe.

H5N1 will probably arrive in North America some day. It's not here now, so all birds are safe from the virus—though they are not necessarily safe from government mass-culling policies.

H5N1 is lethal to chickens and mute swans, less so to ducks. Chickens raised in commercial poultry operations, with large numbers of a single breed confined indoors, live in the ideal conditions for spread of any influenza. Poultry handlers have been among the human victims.

Low Pathogenic Avian Influenza is common in poultry. Birds that live in small flocks tend to catch it, get sick, and get better. Whether they acquire any immunity to other influenzas remains an uninvestigated and unanswered question.

H5N1 has been around for years, possibly as many as forty. It came to attention in 1997 when it started killing chickens in Southeast Asia. The government response was to kill all the chickens within a certain range of a confirmed case of H5N1, the theory being that if there weren't any birds to catch it or pass it on, the virus would be stopped.

The reality of that policy is that not all birds get killed. In countries harboring feral populations, birds evade capture. Rural farmers conceal their birds. Government workers charged with this onerous task don't always follow through with biologically secure methods of carcass disposal.

Vaccination against H5N1 was approved in 2005 for some countries in the European Union and Southeast Asia.

Andrea Heesters

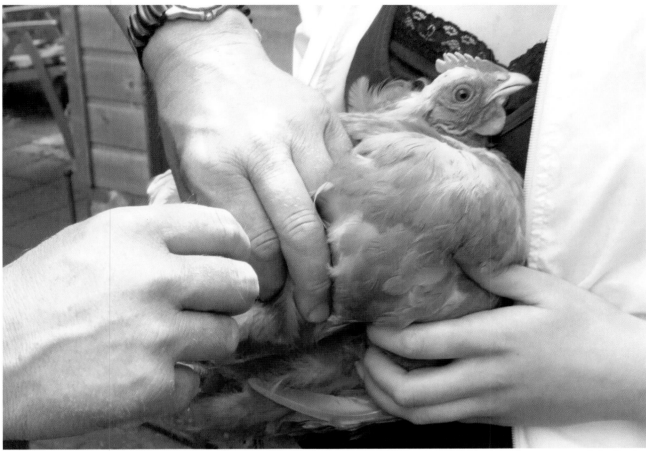

Blood samples are first taken to establish antibody status before vaccination. This bantam Buff Orpington was the bird sampled for the flock.

The vaccination is given into the large breast muscle. This bantam Black Orpington handles the experience calmly. A second dose is required for full protection. Vaccinated birds are exempt from culling in the event of an outbreak of H5N1.

Biosecurity

By following biosecurity practices to create a biological barrier between your birds and the rest of the world, you arm yourself with arguments to establish the safety of your birds and defend them against mass culling. Protect your birds by:

- Participating in the National Poultry Improvement Plan to prevent egg-transmitted diseases. Vaccinate against the diseases your birds are likely to encounter from wild birds and other chickens in your area.
- Maximizing health by not crowding birds and providing lots of fresh air.
- Keeping records of all birds coming in and out.
- Quarantining new birds and birds that have gone to shows for two to four weeks.
- Rotating pastures to prevent parasite and/or disease cycles.
- Removing and quarantining sick birds and any that may have been exposed to them.
- Keeping cats to help control vermin and their parasites.
- Having designated farm clothes that don't leave the farm and are only used for chores.
- Washing any clothes that are worn to any chicken event or tour.
- Removing swallow and pigeon nests from poultry areas and preventing wild ducks and geese from mixing with your birds.
- Minimizing mosquito-breeding habitat on the farm.
- Keeping up-to-date on bird-disease issues.
- Creating and maintaining a visitor log.
- Providing disposable booties or rubber boots to all farm visitors.
- Keeping records of all the above to document your practices.

Trucks used to transport dead birds may not be disinfected, making them carriers of the disease they are intended to stop.

Occasionally, a person catches the flu from a bird. This is rare, because H5N1 is species-specific. Since 1997, more than two hundred cases have been documented, with more than one hundred deaths. Although these figures appear to show a high lethality in humans, the serology studies to identify antibodies have not been done. A study in *The Archives of Internal Medicine* suggests more humans in Vietnam may have been infected and recovered than have been reported.

Many fear that this virus will mutate into a lethal form easily transmissible between humans and swirl around the globe, leaving devastation in its wake.

Gina Kolata's *Flu: The Story of the Great Influenza Pandemic* (2001) awakened interest in that devastating worldwide event. She was one of the *New York Times* reporters covering the 2006 outbreak of avian influenza.

Experts are divided on the degree of threat the H5N1 virus presents for a human pandemic. The *Times* handled it with separate stories on each point of view and a Q&A story by Denise Grady and Kolata, "How Serious Is the Risk?"

Dr. David Nabarro, chief avian influenza coordinator for the United Nations, describes himself as "scared" of the current H5N1 virus. He and the World Bank have raised the specter of 5 million to 150 million deaths, the latter of which is the figure used in headlines. His prediction of disaster has been correct in some cases (AIDS in Africa) to overly dire in others (cholera and malaria in Southeast Asia). He has been a successful fundraiser for the fight against H5N1, collecting $1.9 billion at a summit in Beijing where he hoped to get $1.2 billion.

Dr. Jeremy Farrar of the Hospital for Tropical Diseases in Ho Chi Minh City has treated more actual human H5N1 cases than any other doctor. He noted in the *Times* article that the ratio of conferences to actual cases of the disease is about ten to one. He considers it an unlikely virus to make the jump to transmissibility between humans.

In 2005, reports circulated that H5N1 was causing major bird die-offs in China and Mongolia. Quinhai Lake

Nature Reserve in China reported the deaths of six thousand birds. The lake has many commercial poultry operations on its shores, as well as a fish farm that may use chicken manure as fertilizer and feed. The *Times* stories clarified the reports circulating about these die-offs, establishing that they are either unsupported or far less devastating than rumored. Foreign teams were not allowed to investigate the rumored deaths at Quinghai Lake, and Chinese sampling data were not complete on the species affected. The Wildlife Conservation Society was allowed to send veterinarians to investigate reports of die-offs at Erkhel Lake in Mongolia, where they found only 100 dead birds out of 6,500 of 55 species. Of the 800 samples they collected, only one showed H5N1.

"It has a very low impact," said Dr. William Karesh, one of the veterinarians quoted in the *Times*. "The disease is self-limiting in wild birds."

Effective vaccines for poultry exist. Vietnam, China, the Netherlands, and France have begun vaccinating commercial birds and recommending vaccination for small flocks. Vaccination is controversial because the most commonly used test for H5N1 is an antibody test, which means vaccinated chickens test positive. Since regulations are based on test results, vaccinated birds are a problem for regulations designed to monitor and eliminate avian influenza. Policies continue to evolve. The EU has exempted vaccinated birds from mandatory culling in the event of an H5N1 outbreak. Vaccinated birds may be entered in shows. Vaccination is increasing as rules change to allow birds to be taken to the veterinarian for vaccination rather than requiring the vet to visit the farm.

New methods of testing, such as the rapid polymerase chain reaction tests, can differentiate between infected and

Courtesy of the USDA

David Suarez, veterinary medical officer, and Suzanne DeBois, biological laboratory technician, prepare to load chicken tracheal swab samples into a real-time polymerase chain-reaction machine.

immunized birds, but it is not in general use for poultry yet. It is available for some testing in humans.

Current vaccines for H5N1 in poultry require individual injection, a disadvantage in commercial operations. Owners of small flocks are often willing, even eager, to immunize their birds, but the vaccine is restricted by the government and is not available to them.

Researchers at the University of Pittsburgh School of Medicine led by Andrea Gambotto, assistant professor in the departments of surgery and molecular genetics, have developed a recombinant vaccine, one carrying only the immunity-stimulating properties. Because it can be grown in cell cultures rather than fertilized chicken eggs, it is much faster to produce. They plan a small clinical trial in humans.

In the wake of devastating culling episodes, Vietnam and China began vaccinating, despite the trade problems vaccinating created. It has significantly curtailed the outbreak.

Designation H5N1

The designation H5N1 comes from the way influenzas are classified, first by their internal proteins and then by their surface proteins. The highly pathogenic Asian strain of avian influenza that has dominated the news is an influenza A. A-type influenzas have a segmented internal structure that makes them inherently unstable, resulting in frequent mutations. That's why specific strains are difficult to predict, slowing vaccine development until one strain emerges as the dominant candidate. Influenza A annually kills 36,000 people in the United States according to the Center for Disease Control.

The surface proteins are the antigenic characteristics. The H indicates a hemagglutinin protein, of which there are sixteen. The N indicates a neuraminidase protein, of which there are five. These proteins are on the surface of the virus. That makes 144 possible combinations. Most are host-specific and do not cross species.

Public attention in the United States has focused on the birds in our midst: small flocks, backyard chickens, caged birds, fighting cocks, migratory birds, and others. Scientific investigation has turned up little H5N1 in those groups, but they are frequently cited as carriers and hosts. These groups share a common lack of effective voices speaking in their defense. *The Lancet Infectious Diseases*, an English medical journal, notes that little of the funding pledged to China's Quinghai Lake is dedicated to education or compensation to farmers for dead birds.

The disease is transmitted by sneezing, coughing, and touching and can pass rapidly through birds confined in close quarters at commercial operations. The crowded conditions of commercial poultry operations are also ideal for the mutation of Low Pathogenic Influenza into Highly Pathogenic forms.

The assumption is that these migratory Typhoid Marys of the bird world will carry the infection far and wide. Thus far, the actual transmission routes don't follow that pattern. GRAIN, an international nongovernment organization promoting sustainable management and agricultural biodiversity, and BirdLife International, a global partnership of conservation organizations, offer research that shows H5N1 spreading along major roads and rail lines, following commercial distribution rather than migratory routes.

The 2006 Nigerian outbreak first infected commercial poultry operations in the arid north, leaving waterfowl in the river deltas of the south unaffected. The Nigerian outbreak is currently being attributed to imported day-old chicks from China or Turkey.

In much of the world, small-scale poultry operations are an important part of the economy for the rural poor, providing nearly a third of the protein in their diet. GRAIN issued a report on the effects on rural economies. In 2006, Egypt effectively destroyed its multi-billion-dollar poultry industry and the livelihoods of millions of

Egyptians in a military-style culling operation. In Vietnam, where the average flock size is sixteen birds and half the households—both rural and urban—keep chickens, chickens and eggs supply significant protein to the diet. Joachim Otte, coordinator of the Pro-Poor Livestock Policy Initiative, wrote "It is necessary to very carefully assess HPAI control measures for their differential socio-economic impact across the diverse strata of producers and consumers, weighing potential risks against real livelihoods."

Countries that have reported even a single confirmed case of H5N1 have experienced huge declines in poultry consumption. Gary Butcher, professor of poultry diseases at the University of Florida, has traveled through Asia and the Mideast consulting on H5N1. He has observed people avoiding poultry even in countries that have not reported any cases of the virus.

The virus is not likely to enter the food chain, as commercial production facilities test for even Low Pathogenic Avian Influenza. H5N1 makes chickens so sick that it would be impossible for it to go unnoticed. The virus does not survive long without a live host. Cooking kills it, as does hand washing.

Books are appearing on the subject. Dr. Marc Siegel, associate professor at New York University's School of Medicine and a practicing internist, followed his 2005 book, *False Alarm: The Truth about the Epidemic of Fear*, with *Bird Flu: Everything You Need to Know about the Next Pandemic*, complete with bibliography, in 2006. Lindsey Hillesheim has written *Dead Birds Don't Fly: An Avian Flu Primer for Small-Scale Farmers* (2006) for the Institute for Agriculture and Trade Policy.

This story continues to develop as both breaking news and as a human-interest saga involving the commercial poultry industry, ethnic differences, rural life, Third-World poverty, and preparedness for a disease pandemic and other homeland security issues. Mass culling may give more comfort to the fearful than can be justified by its limited effect on controlling H5N1. While no one is in favor of a global pandemic, informed scientific and political opinions differ on just what threat H5N1 poses.

Dire predictions capture headlines, but public policy based on rumors and fear is taking a serious toll on rural

Andrea Heesters

This goose gets her leg band. The immunization is effective for one year. The pilot program in the Netherlands will be evaluated and recommendations made for future vaccination programs.

lifestyles and Third-World subsistence farm economies. H5N1 has focused attention on the birds around us as updated terrorists secretly setting a bomb that is about to go off. When public policy is directed by the impulse to annihilate this silent enemy among us, collateral damage to rural lifestyles and Third-World farm economies can be worse than the threat.

CHAPTER 14

LEGAL ASPECTS

As with any agricultural enterprise in our complex world, raising chickens is subject to a variety of laws and regulations. Finding out what they are and abiding by them will save you many headaches.

Raising chickens is generally governed by local zoning and land-use laws and ordinances. With the advent of the NAIS and individual state systems, you may also have to register your premises with the state and identify every chicken you own.

Local law may not be the only standard you have to meet. Some planned-unit developments place restrictions on the deeds that prohibit livestock, which is generally understood, if not specified, to include chickens.

Keeping land in agricultural use provides green space. Small farms, under pressure from development as cities expand, may be able to present economic value if they are producing eggs and meat for local markets. Planners are learning the value of green space and may be amenable to persuasion of the importance of small farms.

LOCAL POULTRY ORDINANCES

The laws regulating chickens will be part of the municipal or community laws that govern your property. That could be a city ordinance or, if you live in a rural area, it could be the county policy. Your local health department may also

Courtesy of USDA

Small- and medium-sized diversified farms struggle against the pressures of development. Farming is financially difficult. Rising land values make it tempting to sell to the highest bidder. This country road in Fulton County, Ohio, passes through farmland that is holding out for agriculture.

Chickens such as this White Leghorn conjure rural-life associations that are not always welcome in town. Neighbors may object and local ordinances may restrict poultry.

Courtesy of My Pet Chicken, LLC

have a role in regulating the raising of poultry. In addition, your deed may have restrictive covenants that include chickens. Laws generally regulate the number of chickens permitted, crowing roosters, waste, smell, housing, and backyard butchering.

Chickens are usually classified as livestock or barnyard animals. That makes some people think of them as unsuited to urban and suburban life. Negative attitudes about the status and connotations of keeping chickens in the backyard have resulted in restrictions on them and outright bans.

Locate accurate information from your local government. Most local governments have online resources for those who have Internet access. All local governments are required to provide information to the citizens who live within their boundaries. In some states, municipal authority to control livestock and poultry may extend beyond the municipal boundary. Start with the information desk in the municipal building. Keep asking questions until you find a local government employee who is knowledgeable. As a general rule, erroneous information

provided by a local government official is not a defense for violation of a local ordinance.

Where a restrictive covenant in a deed regulates poultry, local government officials will not have any information about this. You should read your deed and consult an attorney if the language that it contains is difficult to understand. If your neighborhood has a homeowners' association, it may have adopted rules governing the raising of livestock and poultry.

Extension agents, 4-H leaders, and high-school agriculture teachers are good resources for getting started, but they may not know exactly what applies to your property. Check the original documents to make sure you know what you have to do to comply with the law.

If the law is unclear or confusing, ask for help. If no one knows for sure about chickens, pursue the question further. The animal control officer may know how the community deals with chickens and chicken coops, or be able to direct you to someone who does.

If no one, including your elected representatives, can determine what is meant by the law, you may be the person

Madison's Chicken Ordinance

The main restrictions of the chicken ordinance in Madison, Wisconsin, are:

- Not more than four hens per single-family dwelling.
- No roosters.
- Chickens are not allowed to run free. They must be securely enclosed, and the enclosure must be at least twenty-five feet from neighbors' residences.
- No slaughtering on the premises.
- Owners are required to purchase a permit from the city for six dollars a year.

Chicken owners are comfortable abiding by these rules. Chicken ownership in the city has certainly increased since the law was passed in 2004.

Mad City Chickens has become an organized group since the efforts to pass the chicken ordinance. They hold classes in chicken management and building chicken coops. They meet occasionally for potluck dinners and have organized coop tours to share their expertise.

Seattle, Washington, and Portland, Oregon, already had ordinances on their books, giving them models from which to work. Alderperson Matt Sloan was sympathetic.

After six months of publicity to inform the public, backed with experiences of other communities and the endorsement of Mark Cook, a professor in the Poultry Science Department at the University of Wisconsin, the city Common Council adopted an ordinance specifying the conditions for keeping chickens inside city limits in May 2004.

Being a good neighbor is one of the most important ways to avoid complaints about your chickens. In addition to meeting the basics of local laws, always keep your chickens clean and avoid smells. Replace litter frequently and compost or dispose of manure properly. It is a valuable fertilizer, but make sure you don't create a situation that will offend your neighbors. Information about composting can usually be obtained through your local cooperative extension office. Landscaping can camouflage the chicken house and yard, making it more palatable to neighbors.

Giving neighbors fresh eggs occasionally can win over doubtful neighbors. You can also invite them over to meet the chickens and explain your interest to them. Enthusiastic advocates can influence detractors. They may never want chickens for themselves, but at least they can be convinced to tolerate them in your yard.

Some communities have tightened laws about chickens since the avian influenza scare of 2005. You may need to prepare information to reassure neighbors and local officials that your birds represent no danger to anyone.

A restaurant owner in Long Island has kept half a dozen chickens, including a rooster, since he moved to the area in the 1970s. He prefers the fresh eggs for the eggs Benedict he serves his customers. A new law forbidding farm animals in town was not immediately enforced. Existing businesses are usually protected from new laws by a grandfather clause, allowing the business to continue operating as it has in the past.

---

to lead a movement to change it. This happened in Madison, Wisconsin, where, up to 2004, no laws prohibited chickens, but building a coop for them was not allowed. Chicken-lovers kept chickens in their yards, a sort of open secret. So long as no one complained, there was no problem. Occasionally differences arose among neighbors and a city inspector would come out and talk to the chicken owners, sometimes even telling them they would have to get rid of their chickens.

Sometimes they did, and sometimes they simply moved them down the street to a friend's yard for a few months and then quietly moved them back. They considered themselves the Poultry Underground.

In 2003, Alicia Rheal and Bryan Whiting decided to get the law changed to make chickens legal in Madison.

A homeowner in New Jersey had been keeping chickens for a year before someone complained to the city. The Board of Health investigated and found them in violation of local ordinance. After consulting with a lawyer, the family decided to apply for a zoning variance that would make their chickens legal.

These situations reflect the legal confusion that exists in many places. Chickens trigger strong feelings in some situations. Stay calm, research your options, and be pleasant and polite. Nothing is gained from a shouting contest.

Ultimately, you are going to be part of a community. With tact and goodwill, you can convince the community to include your chickens.

VOLUNTARY AGRICULTURAL DISTRICTS

A recent strategy to keep land in agricultural use is the voluntary agricultural district. On the county level, land already in agricultural use can be shielded from nuisance lawsuits and protected from nonfarm development.

Owners of land currently in agricultural use can agree to be included in a voluntary agricultural district, exempting them from sewer and water assessments that are required for residential and commercial development. They usually have some protection from being sued by new neighbors who don't want to hear the rooster crowing.

Ken Hammond, USDA

This farmer continues to plow his fields in Fulton County, Ohio. Residential development has encroached on farmland so that his farm is now next door to suburban development. Farming and suburbia often conflict over issues such as chemical use, waste disposal, and smells.

The self-contained, sustainable farm still exists in Carroll County, Maryland, but it is the exception. "Farmettes" of small acreage provide an alternative for some but risk development as McMansions.

Tim McCabe, USDA

Ken Hammond, USDA

Farmland may gain some protection by claiming other uses, such as wildlife conservation. Landowner Glen Schuetz post signs on his Illinois farm.

Ken Hammond, USDA

The future is in the background for this farm in Fulton County, Ohio. Residential development increases the value of land, which is worth more growing houses than corn.

The agricultural district can work with insurers to provide coverage for agritourism activities. The district offers a structure through which landowners can advance public education about agriculture. Forming a district can attract supportive infrastructure such as feed stores, processing facilities, and marketing opportunities for your eggs and meat.

In an era of mini-estates encroaching on agricultural land, voluntary agricultural districts can maintain land near urban and suburban areas in agricultural uses that enhance the open green space. They are a tool to support agricultural uses against the pressure of development. If your state has an agricultural district program, information is usually available from your state department of agriculture or your local cooperative extension office.

STATE AND FEDERAL REGULATIONS

The United States Department of Agriculture (USDA) has proposed a system to identify every livestock animal in the country. Individual states, such as Wisconsin and Indiana, have also passed laws requiring all properties that have livestock to register with that state. Find out what laws apply in your state in order to comply with them. These programs have aroused a lot of controversy and resistance, so implementation varies across the nation.

The USDA's National Animal Identification System suggests giving every farm animal a seven-digit number and tracking its every move off the premises on which it lives throughout its life. Animal identification by individual Animal Identification Numbers (AIN) is the second

Corallina Breuer

Crowing roosters annoy neighbors unaccustomed to the sounds of farm life. This Naked Neck rooster lets fly. Roosters cannot be trained not to crow, but some crow more than others.

component of NAIS. The third is animal tracing by reporting animal movement. The goal is for animal health officials to be able to trace back to the source of any disease within 48 hours of its discovery.

Radio-frequency technology in the form of implanted microchips is one possible means of tagging animals. Chickens and other poultry may be allowed to be identified by numbered leg or wing bands. Animals that leave their premises for any reason, such as to go to a show, will be required to have an identification number. Even chickens that never leave their premises are encouraged to be identified.

Databases of identified animals will be kept either through breed organizations, commercial producers, state premises registration, or any other databases the USDA needs to track every animal.

Organizations supporting sustainable agriculture, rare-breeds conservation, and individual rights have rallied their members to oppose NAIS. You can join them by contacting any organizations of which you are already a member to find out what you can do to oppose NAIS.

Farm for Life

Farm for Life was established by lawyer and small-farm owner Mary Zanoni to advocate for a sustainable lifestyle based on local communities. She found herself embroiled in opposing the National Animal Identification System (NAIS) in 2005. She provides regular analyses of the USDA's proposals and has spoken widely on the subject.

Zanoni devotes herself full time to following NAIS and developing counterproposals that will satisfy the USDA's need to secure the safety of the food supply without burdening small producers with overwhelming paperwork. For instance, certified organic producers already have a tracing system in place to ensure compliance with organic standards, and could be excluded from NAIS and premises identification without compromising the USDA's goal.

See the appendix for Zanoni's contact information.

Ken Hammond, USDA

Suburban developments put neighbors close to each other. Community agreement on issues such as animals and noise can become contentious. Pet regulations may be used to allow a few chickens in neighborhoods such as this self-help housing project in Florida.

Corallina Breuer

Small-flock owners can find legal ways to pursue their hobby, business, and passion of breeding and raising chickens. This Silkie couple will continue to give pleasure to owners and visitors, sharing their proud ancestry and heritage.

179

AFTERWORD

The future for small flocks is uncertain. Government and industry, responding to consumer demand and political reality, hold power over regulations that could make owning a small flock of chickens nearly impossible. Growing interest in local food sources, animal welfare, and the cost of a centralized food system support small flock ownership. Where they will find common ground and how these influences will find balance is difficult to predict.

Media coverage of avian influenza has taught the public to fear chickens. As better information restores public confidence, the suspicion and anxiety should subside.

Agritourism is encouraging families to explore and understand our agricultural roots. People are hungry for knowledge about their food and the systems that feed them.

Rising fuel costs may do more to encourage small-flock keeping than any other factor. The centralized industrial system that supplies supermarkets and fast-food franchises may become less competitive due to the added cost of transporting animals to processing and meat to the consumer.

Establishment of one or more Fowl Trust Sites could provide an invaluable resource to protect against complete loss of these already-rare breeds. The SPPA has plans to create such a facility.

Ideally, a site of 250 acres or more would allow sufficient space for two hundred or more separate breeds and varieties. Interpretive facilities would help the public understand the contribution and importance of these breeds to our history and future.

Such a resource would be available to rescue endangered flocks, improve breeding stock, and make birds available to new and established flock owners. It could be a repository for historical equipment and publications. Much antique equipment is now in private collections, preserved by those who recognize its value in a disposable world. The solid husbandry that maintained the historic breeds and the ingenuity that created early incubators and other technology remind us of how we need to solve the problems that confront us today.

A library of historic publications would preserve the knowledge gained from lifetimes of breeding and observation. SPPA is seeking records of early poultry clubs, old poultry books and periodicals, and the notebooks of breeders to learn how they managed and what they accomplished. A Fowl Trust Site could serve as a central location for regional and national resources.

Such sites could power the growing interest in small-flock ownership. With more sustainable farms and backyard flocks, breeds now in danger of disappearing would fill show cages as well as fulfilling their historic utilitarian uses. Breeders would vie for top honors among many entries, and novices would have their choice of suitable breeds.

This Golden Laced Wyandotte's piercing gaze looks to the heart of the matter. Chickens see well, using their eyes to locate worms, grubs, seeds, and other foods. Advice on successful flock management is available in books and over the Internet.

Shows would attract huge crowds, and winning breeders find themselves inundated with requests for interviews from *People* magazine. Champion birds would appear on Oprah.

Improved understanding of disease transmission and prevention will open international exchange of chickens. North American, European, Asian, and even world poultry shows will be held.

Establishing the first Fowl Trust Site is a step toward a future filled with chickens. SPPA is a 501c(3) organization able to accept donations of all kinds. Money can purchase land, but a farm with its own history would be an excellent site. If you or someone you know is eager to preserve a family farm that will otherwise be developed, or you can help financially or in kind, contact the SPPA. We are all part of a wonderful future.

APPENDIX

• •

RESOURCES

AGRICULTURAL ORGANIZATIONS

4-H

7100 Connecticut Avenue
Chevy Chase, MD 20815
(301) 961-2800
www.4husa.org
4-H is a part of the Cooperative Extension System, a non-profit program operated through each state's land grant university. The Extension System's staff operates 4-H offices throughout the counties of each state.

ATTRA

National Sustainable Agriculture Information Service
P.O. Box 3657
Fayetteville, AR 72702
Orders: (800) 346-9140
www.attra.org
The National Sustainable Agriculture Information Service is created and managed by the National Center for Appropriate Technology and is funded under a grant from the USDA's Rural Business-Cooperative Service. It provides information and other technical assistance to farmers, ranchers, extension agents, educators, and others involved in sustainable agriculture in the United States.

Farm for Life

P.O. Box 501
Canton, NY 13617
(315) 386-3199
mlz@slic.com
Farm for Life advocates for sustainable lifestyles based on local communities. Its motto is: Farming in harmony, with human life, physical and spiritual, with the life of our Earth, with the life of our fellow creatures. Founder Mary Zanoni publishes a newsletter, $25 a year.

National FFA Organization

P.O. Box 68960
6060 FFA Drive
Indianapolis, IN 46268
(317) 802-6060
www.ffa.org
The National FFA Organization was organized as the Future Farmers of America in 1928 in Kansas City, Missouri. In 1988, the official organization name was changed to the National FFA Organization to reflect the broadening field of agriculture, which today encompasses more than three hundred careers in everything from agriscience to biotechnology to turf grass management. In 1950, Congress granted FFA a federal charter, making it an integral, intracurricular part of public agricultural instruction under the National Vocational Education Acts. Two of the FFA top three executives are employed by the U.S. Department of Education.

POULTRY ORGANIZATIONS

American Bantam Association

P.O. Box 127
Augusta, NJ 07822
(973) 383-8633
www.bantamclub.com

American Pastured Poultry Producers Association

36475 Norton Creek Road
Blodgett, OR 97326
(541) 453-4557
www.apppa.org

American Poultry Association
www.amerpoultryassn.com

Society for the Preservation of Poultry Antiquities
Dr. Charles Everett, Secretary/Treasurer
1057 Nick Watts Road
Lugoff, SC 29078
www.feathersite.com

BREEDS CONSERVATION

American Livestock Breeds Conservancy
P.O. Box 477
Pittsboro, NC 27312
(919) 542-5704
www.albc-usa.org

Rare Breeds Canada National Office
1-341 Clarkson Road
RR1, Castleton, ON K0K1M0
(905) 344-7768
www.rarebreedscanada.ca

HELPFUL BOOKS

Beeton, Samuel Orchart. *Poultry & Pigeons: How to Rear and Manage Them, in Sickness and in Health,* with thirty-eight engravings from designs by Harrison Weir, and two colored illustrations from designs by Harrison Weir and R. Hutulla. London: S. O. Beeton, 1870.

Brown, Edward. *Races of Domestic Poultry.* London: Arnold, 1906.

Damerow, Gail. *Barnyard in Your Backyard: A Beginner's Guide to Raising Chickens, Ducks, Geese, Rabbits, Goats, Sheep, and Cows.* North Adams, MA: Storey Publishing, LLC, 2002.

———. *Building Chicken Coops: Storey Country Wisdom Bulletin A-224.* North Adams, MA: Storey Publishing, LLC, 1999.

———. *Storey's Guide to Raising Chickens: Care / Feeding / Facilities.* North Adams, MA: Storey Publishing, LLC, rev. ed., 1995.

———. *The Chicken Health Handbook.* North Adams, MA: Storey Publishing, LLC, 1994

———. *Your Chickens.* North Adams, MA: Storey Publishing, LLC, 1993.

Davis, P. D. C. and A. A. Dent. *Animals That Changed the World,* adapted from Zeuner, F. E. *History of Domesticated Animals,* with 25 drawings by Sven Berlin. London: Phoenix House, 1966.

Dohner, Janet Vorwald. *Historic and Endangered Livestock and Poultry Breeds.* New Haven, CT: Yale University Press, 2001.

Green-Armytage, Stephen. *Extra Extraordinary Chickens.* New York: Harry N. Abrams, 2005.

Green-Armytage, Stephen. *Extraordinary Chickens,* New York: Harry N. Abrams, 2005.

Hyams, Edward. *Animals in the Service of Man: 10,000 Years of Domestication.* London: Dent, 1972.

Kilarski, Barbara. *Keep Chickens! Tending Small Flocks in Cities, Suburbs, and Other Small Spaces.* North Adams, MA: Storey Publishing, LLC, 2003.

Lee, Andy and Pat Foreman. *Chicken Tractor: The Permaculture Guide to Happy Hens and Healthy Soil.* Buena Vista, VA: Good Earth Publications, 1998.

Lind, L. R.. *Aldrovandi on Chickens. The Ornithology of Ulisse Aldrovandi,* (1600) volume II, book XIV, translated from the Latin with introduction, contents, and notes. 1st American ed. Norman, OK: University of Oklahoma Press, 1963.

Luttman, Gail and Rick. *Chickens in Your Backyard: A Beginner's Guide.* Emmaus, PA: Rodale Books, 1976.

National Geographic Magazine; v. 51, no. 4. *The Races of Domestic Fowl. Fowl of the Old and New World. America's Debt to the Hen.* Washington, D.C.: National Geographic Society, April 1927.

Pangman, Judy. *Chicken Coops: 45 Building Plans for Housing Your Flock.* North Adams, MA: Storey Publishing, LLC, 2006.

Paul, Johannes, William Windham, and Joe Stahlkuppe. *Keeping Pet Chickens: Bring Your Backyard to Life and Enjoy the Bounty of Fresh Eggs from Your Own Small Flock of Happy Hens*. Hauppauge, NY: Barron's Educational Series, 2005.

Percy, Pam. *The Complete Chicken*. Stillwater, MN: Voyageur Press, 2002.

———. *The Field Guide to Chickens*. Stillwater, MN: Voyageur Press, 2006.

Plamondon, Robert. *Success with Baby Chicks*. Blodgett, OR: Norton Creek Press, 2003.

Porter, Valerie. *Domestic and Ornamental Fowl*. London: Pelham Books/Stephen Greene Press, 1989.

Price, Edward O. *Animal Domestication and Behavior*. Wallingford, UK; New York: CABI Pub., 2002.

Rossier, Jay, American Poultry Association, and Geoff Hansen. *Living with Chickens: Everything You Need to Know to Raise Your Own Backyard Flock*. Guildford, CT: The Lyons Press, 2004.

Schmudde, Horst W. *Oriental Gamefowl*. Bloomington, IN: AuthorHouse, 2005.

Smith, H. Easom. *Modern Poultry Development: A History of Domestic Poultry Keeping*. Midhurst, England: Spur Publications Co., c/o Beech Publishing, 1976.

Staples, Tamara, photographs; Ira Glass, essay; Christa Velbel, text. *The Fairest Fowl: Portraits of Championship Chickens*. San Francisco, CA:. Chronicle Books, 2001.

Weir, Harrison F.R.S. *The Poultry Book*. American ed., edited by Johnson, Willis Grant assisted by Brown, George O. as associate editor and many American experts. Complete in eighteen parts. New York: Doubleday, Page & Company, 1904-05.

HELPFUL WEBSITES

Association of Avian Veterinarians. Advances avian medicine and stewardship and refers small-flock keepers to veterinarians who handle chickens. www.aav.org.

BirdLife International. A global alliance of conservation organizations working together for the world's birds and people. www.birdlife.net.

Eggbid. Auction and sale site for birds and hatching eggs. www.eggbid.com.

FeatherSite. Provides information, photographs, and video of chickens and other fowl. www.feathersite.com.

GRAIN. An international nongovernmental organization that promotes the sustainable management and use of agricultural biodiversity. www.grain.org.

Institute for Agriculture and Trade Policy. Aims to create environmentally and economically sustainable rural communities and regions through sound agriculture and trade policy. www.iatp.org.

Mad City Chickens. Gives advice for raising backyard chickens in Madison, Wisconsin. www.madcitychickens.com.

My Pet Chicken. Sells chicken coops, starter kits, and supplies and provides information and resources for raising chickens. www.mypetchicken.com.

Shagbark Bantams. Offers medical and scientific information about shagbark bantams. www.shagbarkbantams.com.

Sproutpeople. Sells over seventy varieties of organic sprouting seeds and sprout supplies. www.sproutpeople.com.

Tarazod Productions. Independent film company run by Robert Lughai and Tashai Lovington that chronicles how backyard chicken owners in Madison, Wisconsin, came out of hiding to get laws changed to allow them to keep chickens legally in the thirty-minute film *Mad City Chickens*. www.tarazod.com.

United States Department of Agriculture. Offers information a variety of information about agriculture production, including how to produce organic meat or eggs, where to find a farmers' market in your area, and how to comply with the National Animal Identification System. www.usda.gov.

AVIAN INFLUENZA RESOURCES

Centers for Disease Control and Prevention, Department of Health and Human Services. Provides information about avian influenza. www.cdc.gov/flu/avian.

Council for Agricultural Science and Technology. Communicates science-based information on food, fiber, agriculture, and natural resources. www.cast-science.org.

Andrew Zimmerman

Many resources are available at low cost or no charge to help you build simple but sturdy chicken enclosures like this chicken tractor.

Food and Agriculture Organization of the United Nations. Works to improve agricultural productivity and food security, and to better the living standards of rural populations. www.fao.org.

PandemicFlu.gov. Provides access to federal government information on avian and pandemic influenza. www.pandemicflu.gov.

ProMED-mail, International Society for Infectious Diseases. Provides updates on the incidence of avian influenza and plans to detect cases. www.promedmail.org.

SPECIALTY BREED CLUBS

Because specialty breed clubs depend on volunteers and extend membership across the country and sometimes around the world, paid staff in a permanent office location is a luxury they do not enjoy. Contact information changes as individuals move in and out of club positions. The Internet is often the best way to locate information and contact a club.

Ameraucana Breeders Club
John Blehm, Secretary/Treasurer
4599 Lange Road
Birch Run, MI 48415
abc@ameraucana.org
www.ameraucana.org
American Australorp Breeders
278 County Road Cna
Champion, MI 49814

American Brahma Club
http://groups.msn.com/AmericanBrahmaClub/homepage.msnw

American Brown Leghorn Club
Miki Schrider, Secretary/Treasurer
125 Victory Lane
Siler City, NC 27344
donandmiki@hotmail.com
www.the-coop.org/leghorn/ablc1.html

Chickens and turkeys settle together at Garfield Farm Museum. A flock of historically accurate poultry is part of a living museum. Visitors respond warmly to chickens.

Courtesy of the Garfield Farm Museum

American Buttercup Club
www.geocities.com/americanbuttercupclub/
Julie Cieslak, Secretary/Treasurer
7257 W. 48 Road
Cadillac, MI 49601
(231) 862-3671
americanbuttercupclub@yahoo.com

American Dutch Bantam Society
Mary Hoyt
9365 N. Santa Margarita Road
Atascadero, CA 93422
hoytwoodacre@charter.net
www.dutchbantamsocietyamerica.com

American Game Bantam Club
Larry Bruffee
10 Pine Street
Shelburne Falls, MA 01370
agbclub@yahoo.com
www.freewebs.com/silver-moon-chickens/links.htm

American Langshan Club
Forrest Beauford, Secretary/Treasurer
18077 S. Hwy 88
Claremore, OK 74017
(918) 341-2238
www.geocities.com/langshan_99/AMERICAN_LANG-
SHAN_CLUB.html

American Marans Club
Wade Jeane, Deputy of the Marans Club of France for the
USA
296 Aspen Lake Drive W.
Newnan, GA 30263
(770) 304-9842
wjeane@charter.net
www.americanmaransclub.com

American Silkie Bantam Club
Lydia Webb, Information
lydlo1@msn.com
Sheila Gordon, Secretary/Treasurer
276 East Palo Verde Ave
Palm Springs, CA 92264
(760) 320-5960
sgordonwindsor@earthlink.net
www.americansilkiebantamclub.org

Araucana Club of America
Nancy Utterback, Membership
11683 North 600 W.
Frankton, IN 46044
www.araucana.freehosting.net

Bearded Belgian d' Anver Club of America
Greg Romer, Secretary
7237 California Lane
Okeana, OH 45053

(513) 738-4012
gsromer@msn.com
www.danverclub.com

Belgian d'Uccle and Booted Bantam Club
Kim Theodore, Secretary/Treasurer
9N 100 Percy Road
Maple Park, IL 60151
belgianduccle@earthlink.net
www.belgianduccle.org

Cochins International
Matt McCammon, Membership Chairman
RT #2 Box 98M
Bloomfield, IN 47424
(812) 384-3777
jacmac@bluemarble.net
http://cochinsinternational.cochinsrule.com/

Cubalaya Breeders Club
http://groups.yahoo.com/group/CubalayaBreedersClub/

Dominique Club of America
Tracey Allen
113 Ash Swamp Road
Scarborough, ME 04074
domchickens@gwi.net
www.dominiquechickens.org

The Dorking Club of North America
Phillip Bartz
1269 Perbix Road, Rt. 1
Chapin, IL 62628
(217) 243-9229
Rooster688@hotmail.com

Faverolles Fanciers of America
Dick Boulanger
69 Perry Street
Douglas, MA 01516
faverolles1@aol.com
www.faverollesfanciers.org

International Cornish Bantam Breeders Association
Don & Dar Karasek
2504 State Road 133
Blue River, WI 53518
(608) 537-2734

Japanese Bantam Breeders Association
Ken Lee, Secretary
6100 N. Panda Point
Dunnellon, FL 34433
(352) 795-9836
mixer100@hotmail.com
http://home.columbus.rr.com/jbba/JBBA.html

The Modern Game Bantam Club of America
Bonnie Sallee, Secretary/Treasurer
P.O. Box 697
Pine Grove, CA 95665
jbsallee@volcano.net

National Frizzle Club of America
Glenda Heywood
Box 1647
Easley, SC 29641
(864) 855-0140
frizzlebird@yahoo.com

National Jersey Giant Club
Robert Vaughn
28143 County Road 4
Pequot Lakes, MN 56472
(218) 562-4067

National Langshan Club
Jim Parker
3232 Schooler Road
Cridersville, OH 45806
polishman@watchtv.net
http://groups.msn.com/NationalLangshanClub

National Serama Club and Registry
8501 Flowe Farm Road
Concord, NC 28025
anthony@seramas.com
www.seramaclub.com

North American Hamburg Society

Mary Hoyt, Secretary/Treasurer

9365 N. Santa Margarita Road

Atascadero, CA 93422

(805) 466-3185

hoytwoodacre@charter.net

www.geocities.com/northamericanhamburgsociety

Old English Game Bantam Club of America

Troy Vannoy

526 East Locust

Collinsville, TX 76233

www.bantychicken.com/OEGBCA

Oriental Game Breeders Association

Eve Bundy

P.O. Box 100

Creston, CA 93432

(805) 237-1010

Polish Breeders Club

Jim Parker

RR6 3232 Schooler Road

Cridersville, OH 45806

polishman@watchtv.net

http://groups.msn.com/PolishChickens

Plymouth Rock Fanciers of America

Pat Horstman, Secretary/Treasurer

5 S. Kings Creek Road

Burgettstown, PA 15021

(724) 729-3701

horstcon@yahoo.com

www.crohio.com/rockclub

Rhode Island Red Club of America

Valerie Gadberry

944 E. Sherman

Hutchinson, KS 67501

www.crohio.com/reds

Rosecomb Bantam Federation

Fran Curtis, Secretary

P.O. Box 109

Thomaston, ME 04861

(207) 354-0711

fjcurtis@midcoast.com

www.rosecomb.com/federation

Russian Orloff Club of America

Curtis Flannery

84505 500 W.

Silver Lake, IN 46982

(574) 566-2426

flanfam@kconline.com

www.feathersite.com/Poultry/Clubs/Orloff/OrlClub.html

Serama Council of North America

Jerry Schexnayder, Treasurer

P.O. Box 159

Vacherie, LA 70090

www.seramacouncilofnorthamerica.com

United Orpington Club

Richard Andree, Secretary/Treasurer

105 Johnson Street NE

Brownsdale, MN 55918

(507) 567-2009

rmandree@mchsi.com

www.geocities.com/srp18407/UOC.html

Wyandotte Breeders of America

Dave Lefeber, Secretary Treasurer

8648 Irish Ridge Road

Cassville, WI 53806

(608) 725-2179

dottestuff@yahoo.com

www.crohio.com/wyan

INDEX

ABOUT
THE
AUTHOR

Corallina Breuer

Author Christine Heinrichs holding a Blue Andalusian chicken.

Christine Heinrichs has been writing about poultry for more than ten years. She found herself captivated by the chickens she and her daughter raised and, as a writer, put her experiences into words. She joined the Society for the Preservation of Poultry Antiquities in 1988 and became its publicity director in 2000.

Christine has a degree in journalism from the University of Oregon, making her a Duck. She is a member of the Society of Environmental Journalists and the Society of Professional Journalists.

She has written about business, engineering, golf courses, and law schools as well as poultry and other animals. She teaches Sunday School and watches wild birds when she isn't watching chickens.

Christine lives in Madison, Wisconsin, with her husband, Gordon, who knew what he was in for when they spent their first date watching chicks hatch. She looks forward to Dorkings in her future.

LOOK FOR THESE ADDITIONAL BOOKS FROM
Voyageur Press

Item number 141805:
The Complete Chicken
ISBN 0896587312

Item number 138123:
The Complete Pig
ISBN 0896586472

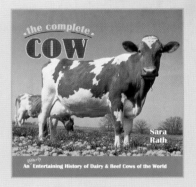

Item number 137215:
The Complete Cow
ISBN 0896580008

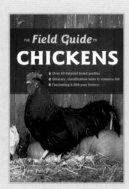

Item number 140590:
The Field Guide to Chickens
ISBN 0760324735

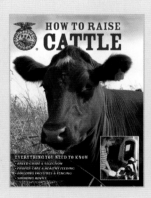

Item number 144215:
How to Raise Cattle
ISBN 0760328021

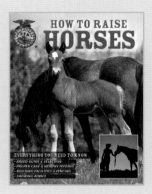

Item number 144139:
How to Raise Horses
ISBN 076032719X

Visit www.voyageurpress.com
or order directly at 1-800-826-6600